Plant Pigments, Flavors and Textures

The Chemistry and Biochemistry of Selected Compounds

FOOD SCIENCE AND TECHNOLOGY

A SERIES OF MONOGRAPHS

A complete list of titles in this series appears at the end of this volume.

Plant Pigments, Flavors and Textures

The Chemistry and Biochemistry of Selected Compounds

N. A. MICHAEL ESKIN

Department of Foods and Nutrition
University of Manitoba
Winnipeg, Manitoba, Canada

ACADEMIC PRESS New York San Francisco London 1979
A Subsidiary of Harcourt Brace Jovanovich, Publishers

ACADEMIC PRESS, INC.
111 Fifth Avenue, New York, New York 10003

United Kingdom Edition published by
ACADEMIC PRESS, INC. (LONDON) LTD.
24/28 Oval Road, London NW1 7DX

Library of Congress Cataloging in Publication Data

Eskin, Neason Akivah Michael.
Plant pigments, flavors and textures.

Includes bibliographical references and indexes.
1. Color of food. 2. Flavor. 3. Food texture.
I. Title.
TX531.E83 664 79-13474
ISBN 0-12-242250-3

PRINTED IN THE UNITED STATES OF AMERICA

79 80 81 82 9 8 7 6 5 4 3 2 1

This book is dedicated to my wife
Nella
and our three sons
Katriel, Joshua and Ezra
and to
the memory of my late father
Rev. Joshua Eskin

"Who is wise? He who learns from all men"
Talmud.

Contents

Preface

The quality of a food is judged by the consumer according to its appearance, taste and texture. This book focuses on the chemistry and biochemistry of compounds responsible for the pigments, flavors, and textures of some fruits and vegetables. Since much of the information presented is scattered in the scientific literature, an attempt has been made to integrate the material into a concise yet comprehensive text.

The subject matter is organized into three sections dealing separately with pigments, flavors and textures. Section I discusses pigment degradation during processing and storage as well as attempts to prevent color deterioration. Section II examines the biogenesis of several groups of compounds that contribute to flavor. Finally, Section III deals with the chemistry and biochemistry of the plant cell wall components and their relation to texture.

During the preparation of this book I have tried to present information that would be useful to food scientists as well as to other's interested in foods. The extensive references cited in the text will enable the reader to pursue any of the topics discussed, in more depth. The discussion of rather select areas will, it is hoped, point out to the researcher where there are still wide gaps in our knowledge.

I am indebted to my wife, Nella, for her assistance, suggestions and encouragement throughout the preparation of this book. I thank CRC Press Inc., for permitting me to use part of an article I wrote in "CRC Critical Reviews in Food Science and Nutrition" vol. 9, pp. 1–40 (1977) for inclusion in Chapter 6. I am also grateful to Professor Marion Vaisey-Genser for her review of several chapters and for suggestions during the preparation of this manuscript.

Plant Pigments, Flavors and Textures

The Chemistry and Biochemistry of Selected Compounds

SECTION I

Food Pigments: Deterioration and Stabilization

Color is recognized as a major factor affecting food acceptance and has been reported to account for approximately 45% of the total quality scale (Kramer and Twigg, 1956). The consumer expects a specific food to possess a well-defined color and rejects any appreciable deviation from the norm. Color deviations accompanied by flavor and textural changes may indicate spoilage or inferior products. The change in color during the preparation of food is a useful index of doneness. Color also provides a useful guide to quality control and is used by many food processors as the criterion for selecting raw materials.

The characteristic color of raw food is due to the natural pigments present in the plant and animal material. These can be enhanced in processed food products through the addition of food colorants. This section discusses pigment changes in raw and processed foods and some of the methods used to retain these pigments in order to increase consumer acceptance. Although these pigments are numerous, they can be classified into three major groups according to their chemical structures: (1) tetrapyrrole derivatives; (2) isoprenoid derivatives; and (3) benzopyran derivatives.

Chapter 1

Tetrapyrrole Derivatives: Chlorophylls

I. INTRODUCTION

Chlorophylls, the green pigments intimately involved in photosynthesis, are located in the chloroplasts of plants. They are extremely complex organic molecules in which a magnesium atom is held at the center of a tetrapyrrole ring. Several derivatives are recognized, although chlorophylls *a* and *b* appear to be the principal ones responsible for the green color of vegetables and are normally present in the ratio 3 : 1. The chemical formulas of these two chlorophyll derivatives are as follows:

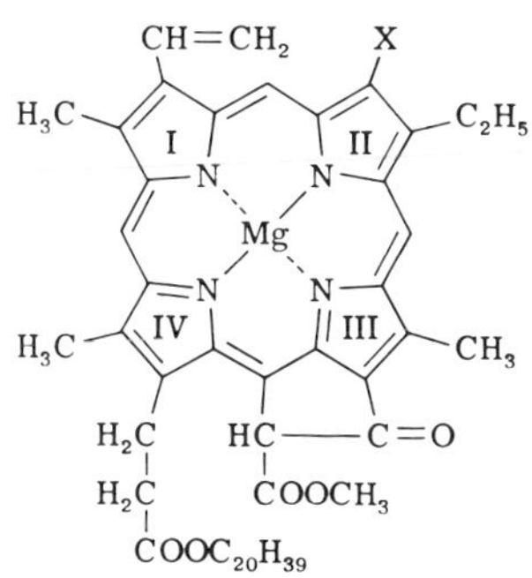

Chlorophyll *a*, X = CH_3

Chlorophyll *b*, X = CHO

The loss of color associated with the processing and storage of green vegetables is a problem of great concern to the food technologist. Increased color retention can result only by improving the stability of the chlorophyll pigments involved. The degradation of these pigments appears to involve one or more of the following reactions:

1. Replacement of the magnesium atom in the chlorophyll molecule by hydrogen to form pheophytin.
2. Removal of the phytol group from the chlorophyll molecule to form chlorophyllide by the enzyme chlorophyllase:

   $$\text{chlorophyll} + H_2O \xrightarrow{\text{chlorophyllase}} \text{chlorophyllide} + \text{phytol}$$

   Following the formation of chlorophyllide, replacement of the magnesium atom by hydrogen can occur, resulting in the formation of pheophorbide.
3. Oxidative reactions leading to the degradation of chlorophyll to colorless products. Several enzymes have been implicated, including lipoxygenase.

II. PHEOPHYTIN FORMATION

The formation of pheophytin in green vegetables is accompanied by an undesirable color change from bright green to a dull olive green. Campbell (1937) recognized this as the reaction responsible for the discoloration of frozen peas stored at −6.9°C. The determination of the amounts of chlorophyll and pheophytin present in frozen vegetables as an index of quality changes was proposed by Mackinney and Weast (1940), who described a method to quantify the ratio of these two compounds. Conversion of chlorophyll to pheophytin in the presence of dilute acids was first reported by Willstater and Hocheder (1907) and was demonstrated by later investigators to proceed in the presence of natural plant acids. Joslyn and MacKinney (1938) found the rate of pheophytin formation to be first order with respect to the acid concentration. Wagenknecht *et al.* (1952) related the color loss in stored frozen raw peas to the increase in free fatty acids while Dietrich *et al.* (1957) suggested the different rates of chlorophyll degradation in peas and beans was due to their differences in acidity. The relationship between appearance and pheophytinization of chlorophyll was examined by Walker (1964a) using frozen green beans stored over a period of 32 months. A linear relationship was evident (Fig.

1.1), although after 12 months of storage other degradative reactions were also thought to contribute to the color changes.

The relationship between chlorophyll degradation and water activity (a_w) was studied by LaJollo and co-workers (1971). They followed the loss of chlorophyll in freeze-dried, blanched spinach purée stored at 37°C and 55°C under nitrogen or air. Water activity had a definite effect, which resulted in a rapid loss of color from bright green to the dull olive green characteristic of pheophytins. Formation of pheophytins at 37°C appeared to be the predominant reaction at a_w levels greater than 0.32. The rate of breakdown of chlorophyll *a* was considerably faster than that of chlorophyll *b* by a factor of 2.5–3.0, which was in agreement with the figure reported by previous investigators, although the factor was found to vary (Schanderl *et al.*, 1962; Gupte et al., 1964). A linear relationship was evident between a_w and the log time for a 20% loss of chlorophyll observed (Fig. 1.2). Under these conditions the chlorophyll/pheophytin

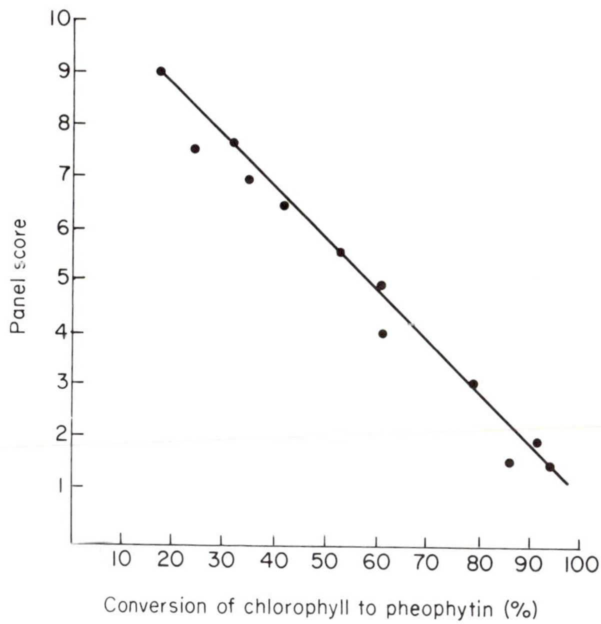

Fig. 1.1. Relationship between the color score of green beans and percent conversion of chlorophyll to pheophytin. Color was assessed by a trained panel using a score of 10 for fresh beans and 0 for extremely yellowing beans. (Walker, 1964a. Copyright © by Institute of Food Technologists.)

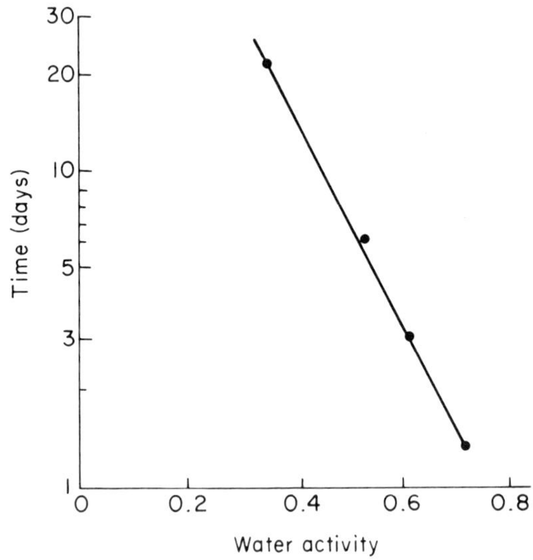

Fig. 1.2. Time required for 20% loss of chlorophyll in spinach at different water activities (37°C in air). (LaJollo *et al.*, 1971. Copyright © by Institute of Food Technologists.)

ratio provided a useful indication of quality, although at a_w levels below 0.32 the rather slow pheophytinization of chlorophyll rendered this relationship impractical.

III. CHLOROPHYLLIDE AND PHEOPHORBIDE FORMATION

The effects of blanching on green vegetables were studied by Jones *et al.* (1963) using okra, turnip greens, and snap beans. These vegetables were blanched at two temperatures, 82.2° and 100°C, for 4 minutes and analyzed for their chlorophyll derivatives (Table 1.1). Considerable formation of chlorophyllides and pheophorbides was evident in okra and turnip greens blanched at 82.2°C, which was in agreement with observations by Mackinney and Weast (1940) that chlorophyllase activity became more marked when some green vegetables were subjected to hot water. The vegetables blanched at the higher temperature showed no chlorophyllide or pheophorbide formation, indicating the complete inactivation of chlorophyllase under these conditions. Buckle and Edwards (1970) reported considerable pheophytinization of chlorophylls in unblanched peas during frozen storage, accompanied by the formation of chlorophyl-

TABLE 1.1

Pigment Content of Unblanched versus Blanched Okra, Turnip Greens, and Snap Beans[a]

Vegetable	Chlorophylls and chlorophyll derivatives[b,c]							
	C_a	C_b	Cd_a	Cd_b	Py_a	Py_b	Po_a	Po_b
Okra								
Unblanched	53.3	27.2	0.0	0.0	15.1	4.4	0.0	0.0
Blanched, 82°C	22.1	14.4	26.5	12.1	13.4	0.5	7.5	3.5
Blanched, 100°C	48.2	25.8	1.3	1.2	18.3	4.2	0.4	0.5
Turnip greens								
Unblanched	62.7	28.1	0.0	0.0	8.6	0.7	0.0	0.0
Blanched, 82°C	51.2	24.8	4.3	0.7	13.0	−0.2	0.3	5.9
Blanched, 100°C	60.5	27.6	0.0	0.0	12.1	−0.2	0.0	0.0
Snap beans								
Unblanched	49.4	25.2	0.0	0.0	17.7	7.7	0.0	0.0
Blanched, 82°C	40.3	24.8	0.0	0.0	27.8	7.1	0.0	0.0
Blanched, 100°C	36.7	23.9	0.0	0.0	29.1	10.2	0.0	0.0

[a] From Jones *et al.* (1963). (Copyright © by Institute of Food Technologists.)
[b] Values expressed as a percentage of total pigments.
[c] Abbreviations: C_a, chlorophyll *a*; C_b, chlorophyll *b*; Cd_a, chlorophyllide *a*; Cd_b, chlorophyllide *b*; Py_a, pheophytin *a*; Py_b, pheophytin *b*; Po_a, pheophorbide *a*; Po_b, pheophorbide *b*.

lides and pheophorbides. This is contrasted with the blanched samples, in which no phytol-free pigments were found. The presence of these phytol-free pigments had not been reported previously in unblanched peas, although the presence of chlorophyllase activity could not be confirmed.

The conversion of chlorophylls to chlorophyllides and pheophorbides during brining of cucumbers was first observed by Jones *et al.* (1961). Their results indicated that brining favored chlorophyllase activity by denaturing the chloroplastin in much the same way that hot water did (Weast and Mackinney, 1940). Further investigations by Jones *et al.* (1963) indicated that pigment changes in whole and sliced cucumbers were induced by the various blanching and brining treatments. Considerably more chlorophyllide and pheophorbide production was found in the unbrined cucumbers blanched at 82.2°C than in those blanched at 100°C, although the small amounts of these derivatives observed at the higher blanching temperature indicated some residual chlorophyllase activity. The corresponding brined cucumbers all exhibited pheophytin and pheophorbide formation. Those blanched at 82.2°C compared favorably with the equivalent unblanched cucumbers in that approximately 80% of the total pigments appeared as pheophorbides. Approximately 25% of the total pigments of brined cucumbers blanched at 100°C were pheophorbides, indicating the incomplete inactivation of chlorophyllase under these conditions. The amount of pheophorbides in the brined cucumber samples was almost twice that reported in the corresponding unbrined cucumbers. It was apparent, therefore, that both blanching and brining were responsible for the formation of chlorophyllides, pheophorbides, and pheophytins. White *et al.* (1963) demonstrated that the conversion of chlorophylls and chlorophyllides to pheophytins and pheophorbides occurred immediately following the commencement of brining. The in-

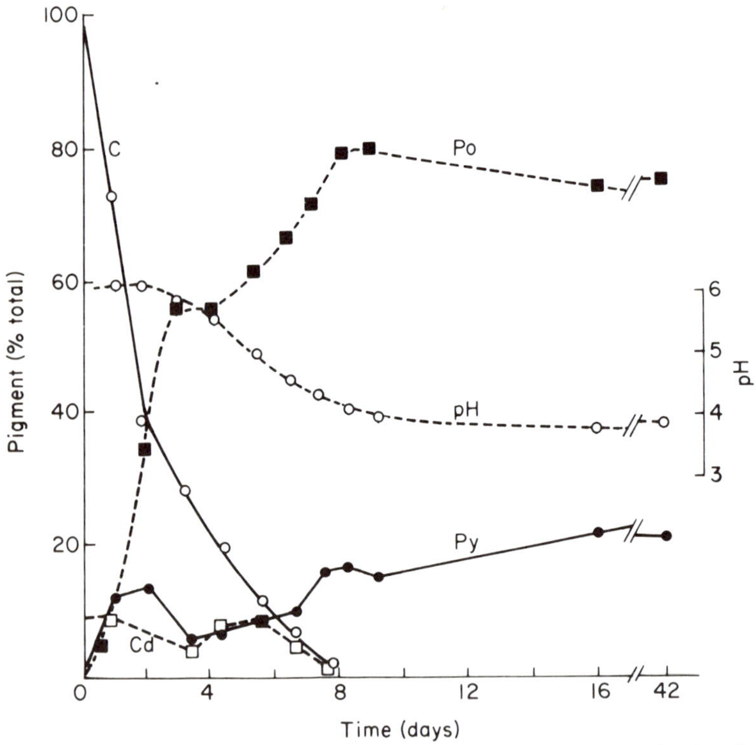

Fig. 1.3. Changes in percentage of chlorophylls (C), chlorophyllides (Cd), pheophytins (Py), and pheophorbides (Po) in brined cucumbers. (White *et al.*, 1963. Copyright © by Institute of Food Technologists.)

crease in acidity of the brine during the course of fermentation was thought to be responsible for the conversion of chlorophylls and chlorophyllides to their corresponding magnesium-free derivatives (Fig. 1.3).

IV. OXIDATIVE CHANGES

A. Enzymatic

Strain (1941) identified an unsaturated fat oxidase in the aqueous extracts of soybeans that was capable of bleaching chlorophylls *a* and *b* in the presence of fat and oxygen. This enzyme, later identified as lipoxygenase, was reported by Mapson and Moustafa (1955) to be present in peas. It catalyzed the oxidation of a number of substrates including chlorophyll when incubated with linoleic or linolenic acid. Lipoxygenase (linoleate oxidoreductase, EC 1.13.11.12) has since been identified in a large number of fruits and vegetables. It is specific for fatty acids possessing *cis,cis*-1,4-pentadiene groups, converting them to optically active *cis,trans*-diene hydroperoxides:

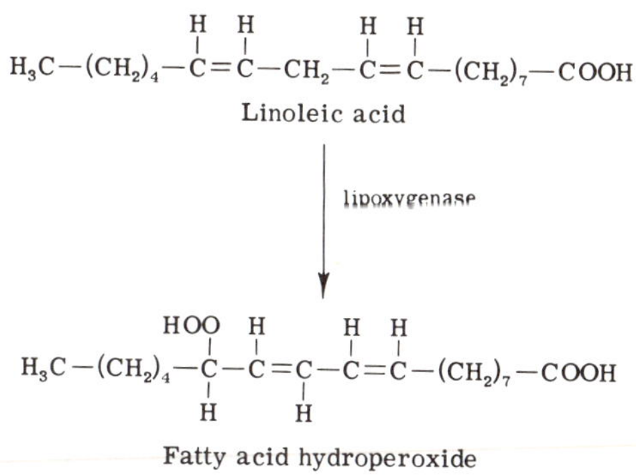

Wagenknecht *et al.* (1952) attributed the loss of chlorophyll in frozen raw peas to a rise in acid number and the activities of the enzymes lipase and lipoxygenase. Further investigations by Wagenknecht and Lee (1956, 1958) confirmed the role of lipoxygenase in both the lipid oxidation and chlorophyll losses observed in green peas. Holden (1965) carried out extensive studies on the chlorophyll bleaching and lipoxygenase activities of a large number of legume seeds including peas and soybeans. She reported that the bleaching of chlorophyll appeared to be coupled with the breakdown of the fatty acid hydroperoxides. The degradation of these hydroperoxides was reported to require a thermolabile bleaching factor present in the crude extracts since no bleaching activity was evident in the presence of purified lipoxygenase preparations. The existence of such a factor was reported by a number of investigators and was thought to break down the linoleate hydroperoxides formed by the action of lipoxygenase (Blain and Barr, 1961; Blain and Styles, 1959; Gardner and Clagett, 1965; Gini and Koch, 1961). Buckle and Edwards (1970), using model systems containing pea and soybean extracts, similarly observed that the loss of chlorophyll and lipid oxidation during storage resulted from the combined action of lipoxygenase and a lipohydroperoxide breakdown factor present in the crude extracts. This factor was identified with the heat-labile factor described earlier by Holden (1965).

Zimmerman (1966) characterized an enzyme from flaxseed, linoleate hydroperoxide isomerase, which catalyzed the isomerization of the unsaturated fatty acid hydroperoxide to the corresponding ketohydroxy compound. Zimmerman and Vick (1970) later identified this enzyme in barley (*Hordeum vulgare*), wheat germ (*Triticum aestivum*), mung bean (*Phaseolus aureus*), and soybean (*Glycine max*). They further reported that chlorophyll bleaching was evident only in the presence of the crude extracts and was not catalyzed by either soybean or flaxseed lipoxygenase preparations. Inactivation of the isomerase enzyme was achieved by heating the extracts at 100°C for 5 minutes, which also resulted in a loss of bleaching activity. This provided confirmatory evidence for the importance of this enzyme in the bleaching process. When isolated from flaxseed it exhibited a pH optimum of 7.0 and was completely inactivated by heating to 68°C for 1 minute.

Zimmerman and Vick (1970) concluded that the heat-labile factor referred to by Holden (1965) was probably identical with the isomerase enzyme. Confirmation of this was provided by the fact that a pH optimum of about 6.0 reported by Holden (1965) for the bleaching of chlorophyll by legume seeds was in close agreement with the pH optima observed for hydroperoxide isomerase in barley, wheat germ. and soybean by Zimmerman and Vick (1970). The bleaching effect was attributed to an oxidation–reduction reaction involving the ketohydroxy fatty acid and a portion of the conjugated bond system of the chlorophyll molecule. In more recent studies, Imamura and Shimizu (1974) were unable to demonstrate the participation of hydroperoxide isomerase in the bleaching of chlorophyll as reported previously by Zimmerman and Vick (1970). These researchers separated lipoxygenase and hydroperoxide isomerase from

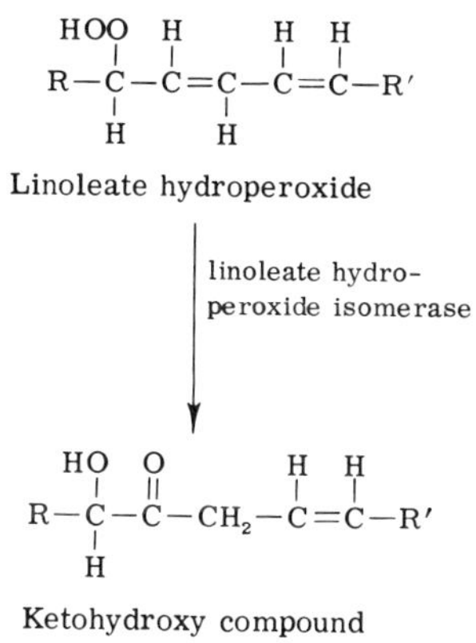

wheat leaves by gel filtration on Sephadex G-200 and found that chlorophyll bleaching activity resided only in the lipoxygenase fraction. Imamura and Shimizu (1974) suggested that multiple lipoxygenases present in plants might promote bleaching. This was evident in the potato tuber, in which two types of lipoxygenases were found, one of which exhibited cholorophyll bleaching activity. However, further clarification is required to determine the precise relationship between the type of lipoxygenase and the bleaching activity associated with different plant sources.

B. Nonenzymatic

Walker (1964b) investigated the effect of blanching on color deterioration in frozen French beans and reported that blanching times greater than 1 minute resulted in a lag period followed by a reduction in chlorophyll. The rate of oxidative degradation was found to increase with storage and longer blanching periods (Fig. 1.4). The effects of blanching appeared to be twofold. First, by inactivating the enzymes involved in chlorophyll degradation, blanching improved the stability of the pigment during storage. Second, although excessive blanching periods inactivated deteriorative enzymes, oxidative reactions were initiated, which resulted in significant chlorophyll losses. Consequently, for maximal color retention an optimal blanching period of 45 seconds to 1 minute was recommended since it prevented both enzymatic activity and heat-initiated nonenzymatic processes.

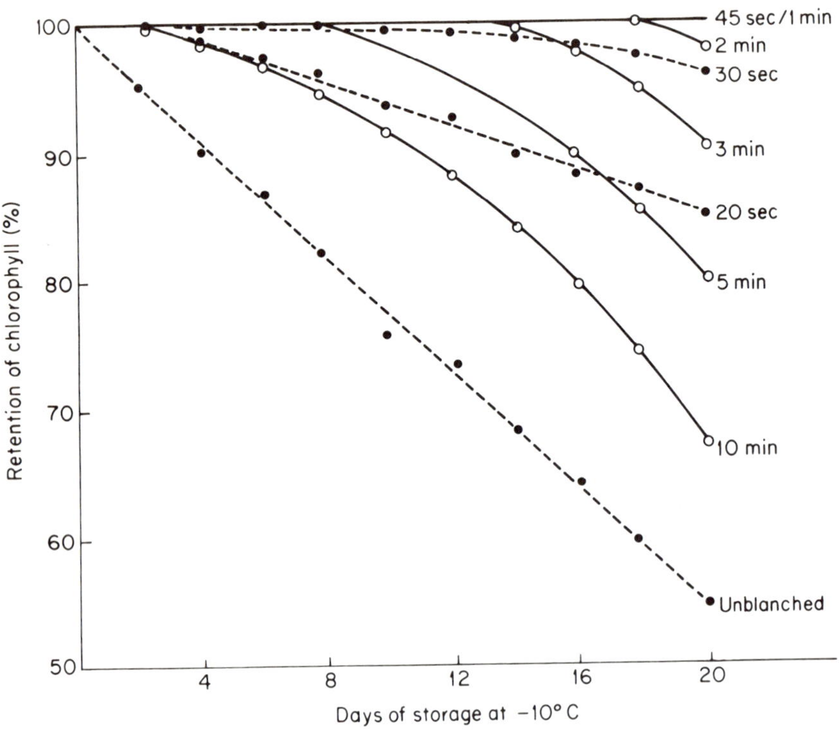

Fig. 1.4. Destruction of chlorophylls during storage at −10°C in green beans blanched for various periods. Dashed lines indicate blanching times of 30 seconds or less (Walker, 1964b).

V. EFFECT OF THERMAL PROCESSING

The loss of color associated with the thermal processing of green vegetables is similarly related to chlorophyll-pheophytin conversions. Gold and Weckel (1959) followed the decline in color during the sterilization of canned peas and attributed the loss of color to these reactions. The formation of pheophytin was correlated with an increase in plant acids during the heating process and subsequent storage. Mori *et al.* (1966) also noted a similar relationship between the rise in total acidity and chlorophyll degradation during the heating process and storage of green vegetables. Tan and Francis (1962) investigated the effect of processing temperature on spinach pigments and noted that color changes were initiated by the formation of pheophytin *a* from chlorophyll *a* followed by the corresponding formation of pheophytin *b*. The degree of conversion appeared to be less marked at the higher processing temperatures, which was verified in later studies by Gupte *et al.* (1964).

VI. REGREENING OF PROCESSED VEGETABLES

A phenomenon occasionally observed in the commercial packing industry is the regreening of processed vegetables. Schanderl *et al.* (1965a) stored asceptically canned green pea purée in the dark at temperatures ranging from 0° to 37°C and found that canned products stored for 6 months at 37°C exhibited a reversion of their green color. This color resembled that of the original unprocessed product but appeared to be far more stable. The pigment responsible was subsequently isolated and found to contain copper and zinc. The reversion effect appeared to involve the complexing of these metals with the breakdown products of chlorophylls *a* and *b*. Schanderl *et al.* (1965a) reported the formation of these complexes when copper was present at 1–2 ppm; 10–20 ppm was sufficient for complete complexation with the pheophytins present in the pea purée. Complexing with zinc, however, did not commence until 25 ppm were present, and 100 ppm were required before complex formation was complete. These investigators concluded that under suitable conditions pheophytins, and in some foods pheophorbides, form green complexes with trace amounts of copper and zinc. These complexes are responsible for the regreening effect and improved color stability. Using model systems, Schanderl *et al.* (1965b) carried out further studies on the formation of these metal complexes and reported that the ratio of metal to pheophytin or pheophorbide was 1 : 1.

Swirski *et al.* (1969) noted that Brussel sprouts picked before the first autumn frosts produced a water-soluble green pigment when heated in water at 115°C. This stable green pigment was attributed to the formation of a complex between a water-soluble protein moiety containing copper and zinc and chlorophyll and its derivatives.

VII. PRESERVATION OF CHLOROPHYLL PIGMENTS

Many attempts to stabilize the green pigments of processed vegetables have been reported, although at present no completely satisfactory method is available. Sweeney and Martin (1961) reported a progressive increase in chlorophyll *a* retention when frozen green beans were cooked in citrate–phosphate buffers pH 6.2–7.0. No further improvement in color was evident when these vegetables were cooked above pH 7.0, although a marked deterioration in flavor was observed. Since an acidic environment is generally regarded as favoring pheophytinization of chlorophyll, elevating the pH of vegetables with approved alkalizing agents was investigated. Several patents were developed using these substances to inhibit the acidic degradation of chlorophyll (Bendix *et al.*, 1952; Blair, 1940a,b; Gieseker, 1949). Although some improvements were evident in the appearance of the vegetable product, color losses were apparent during storage. Of the alkalizing agents recommended, magnesium carbonate was found to be particularly suitable (Gupte *et al*, 1964), although loss of color was still evident during the storage period. Eheart and Odland (1973) investigated the use of ammonium compounds to retard pheophytin formation in frozen green vegetables. Chlorophyll losses were markedly reduced by the use of ammonium hydroxide in microwave blanching of frozen broccoli. A 27% improvement in chlorophyll retention over controls was reported for frozen green beans blanched in ammonium bicarbonate. Although the chlorophyll *a* and total chlorophyll of the treated frozen green beans had not changed significantly after 6 months of storage at $-15°C$, there was a significant loss of chlorophyll *b*.

An alternative approach was developed from an observation by Borodin (1882) that under certain conditions chlorophyll appeared to be more stable. This is now recognized to be due to the conversion of chlorophyll to chlorophyllide by the action of chlorophyllase under these conditions. Thung (1963) reported that chlorophyllides helped to stabilize the desirable color of spinach and attempted to optimize the conditions for its formation. Loef and Thung (1965) later reported some improvement in quality accompanying the formation of chlorophyllides in spinach, the extent of which varied with season. Clydesdale and Francis (1968) tried to control color losses in processed spinach by a combination of pH elevation and preprocess conversion to chlorophyllides. Some improvement in color was reported, although great difficulty was encountered in producing sufficient chlorophyllides. A further attempt was reported by Clydesdale *et al.* (1970) to improve color retention in spinach purée. These investigators added digitonin to spinach purée to release both chlorophyll and chlorophyllase from their association with lipoproteins within the chloroplast. The surfactant exhibited a protective effect on the green pig-

TABLE 1.2

Effect of $MgCO_3$ and/or Digitonin on the Pigment Content of Spinach Puree[a,b]

Pigment	0.35% $MgCO_3$	0.1% digitonin	0.35% $MgCO_3$ + 1% digitonin
$C_a + Cd_a$	52.51	48.50	52.09
$C_b + Cd_b$	16.28	21.20	18.92
$Py_a + Pd_a$	21.20	24.30	21.89
$Py_b + Pd_b$	10.01	6.00	7.10

[a] From Clydesdale *et al.* (1970). (© Food Product Development.)

[b] Values expressed as a percentage of total pigment.

ment similar to that observed with the alkalizing agent magnesium carbonate (Table 1.2) in preventing chlorophyll and chlorophyllide degradation.

The application of high-temperature short-time (HTST) methods in vegetable processing has received considerable attention in recent years. The extensive degradation of chlorophyll pigments associated with conventional heat processing operations was found to be considerably reduced by the use of HTST processing methods (Brody *et al.*, 1960; Joffe *et al.*, 1961; Luh *et al.*, 1964). Tan and Francis (1962) reported that increasing the processing temperature from 115.6° to 137.8°C resulted in a much greater retention of chlorophyll pigments in spinach purée, although rapid degradation was still apparent during storage. Gupte and Francis (1964) investigated the combination of pH adjustment and HTST processing on pigment retention in spinach purée. The improvement in color observed was attributed primarily to a decrease in chlorophyll *a* degradation; however, the product still lacked stability during storage. Buckle and Edwards (1970) studied the changes in chlorophyll, color, and pH during the HTST processing of green pea purée and reported significant correlations between pheophytin production and color measurement. The extent of chlorophyll degradation appeared to be dependent on the processing temperature, product pH, storage time, and storage temperature. Improved retention of the green color was to a large degree determined by the pH changes taking place during the processing and storage periods. These results indicated that controlling pH changes offered the best means of preserving the green color of the processed vegetables. Clydesdale and co-workers (1970) noted that any initial color improvement from HTST processing was lost after 3 months of storage. They attributed this to the fact that the pH of both conventional and HTST processed products exhibited the same trend during storage. In an attempt to stabilize the pH they added magnesium carbonate and phosphate buffer systems but found that this necessitated the addition of unacceptable amounts of the buffer salts. It is evident from these studies that further research is required to establish more effective methods of preventing pigment losses.

Chapter 2

Isoprenoid Derivatives: Carotenoids

I. INTRODUCTION

Many fruit and vegetable colors fall within the pale yellow to purplish-red range. Such pigmentations are determined by two groups of compounds: carotenoids and anthocyanins. Carotenoids are fat-soluble pigments composed of isoprene units, whereas the anthocyanins are related to the flavonoids and are water soluble, as discussed in the next chapter. The isoprene units form a system of conjugated double bonds in the carotenoid molecule, which is responsible for the observed color. The all-*trans* configuration predominates in the majority of these polyenes, providing the most stable structure. Under conditions of processing and storage, however, *cis*–*trans* isomerization occurs.

Carotenoids are generally classified into two groups: carotenes and xanthophylls. The former are structurally related to the hydrocarbons, whereas xanthophylls represent the corresponding hydroxy, epoxy, and oxy derivatives and are frequently esterified. Examples of the first group are α- and β-carotenes, and zeaxanthin is a widely occurring xanthophyll.

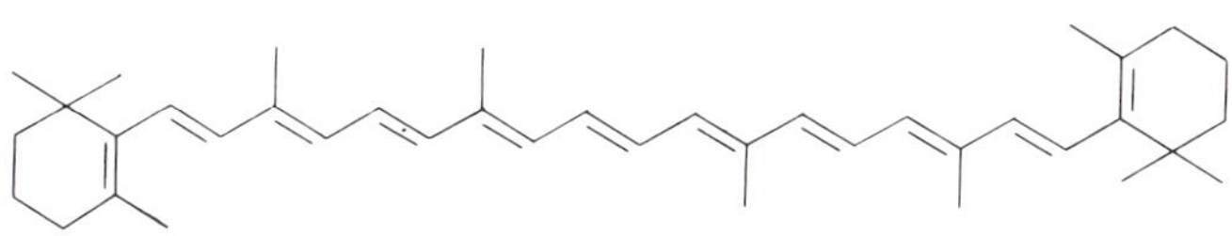

β-Carotene

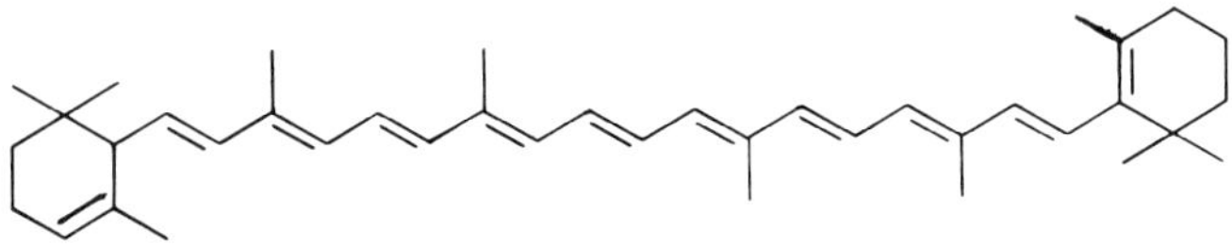

α-Carotene

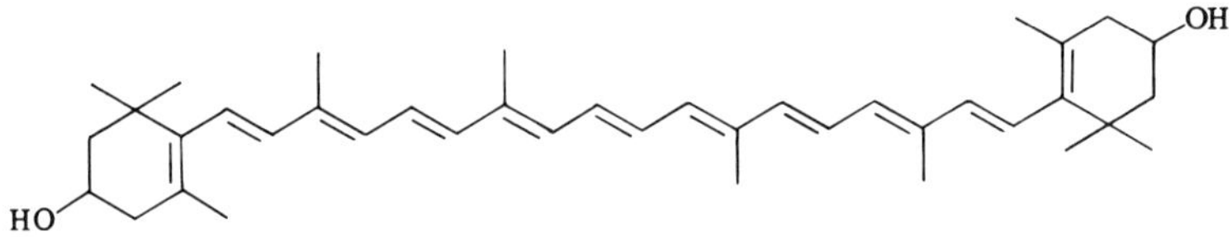

Zeaxanthin

The carotenoids that impart a desirable color to foods are based structurally on α- and β-carotenes and lycopene (the latter pigment is present in the tomato fruit). The stability of carotenoids varies widely in foods, although their unsaturated nature renders them susceptible to oxidation.

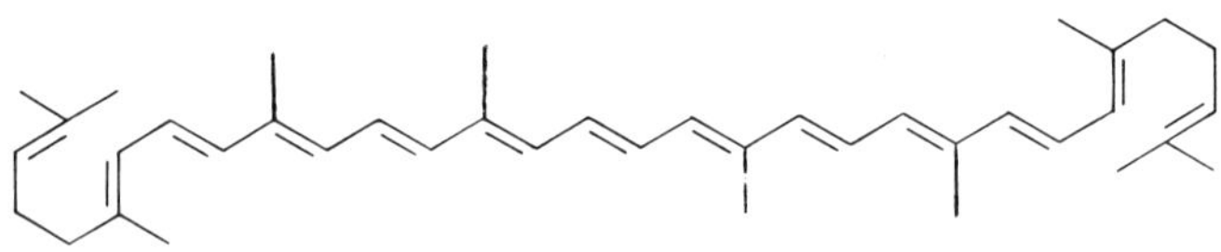

Lycopene

In addition to such oxidative changes, isomerization from the *trans* to *cis* configuration can also result in some degree of color loss. This chapter deals with reactions that are known to have a deleterious effect on the integrity of these pigments in foods with respect to color only and is not concerned with provitamin activity.

II. OXIDATIVE CHANGES

A. Enzymatic

The presence of carotenoid bleaching activity in fresh plant tissue is well established. The responsible enzyme, lipoxygenase (linoleate oxygen oxidoreductase, EC 1.13.11.12), was identified in alfalfa by Sumner (1943), although its activity was considerably less than that reported for soybean. Mitchell and Hauge (1946) demonstrated the acceleration of carotenoid breakdown in alfalfa suspensions by the addition of linoleic acid, thus confirming the role of this enzyme in the bleaching process. Lipoxygenase appears to catalyze the direct oxidation of fatty acids possessing *cis,cis*-1,4-pentadiene groups with the concurrent bleaching of carotenoids, as discussed earlier in relation to chlorophylls. Among the products formed are apocarotenals and a series of carotenoid aldehydes (Blain, 1963; Winterstein *et al.*, 1960). The presence of this enzyme has since been demonstrated in a wide variety of fruits and vegetables including legume seeds, eggplant, cauliflower, and avocado (Pinsky *et al.*, 1967; Rhee and Watts, 1966).

In recent years there has been some confusion regarding the use of the term "carotene oxidase" to describe lipoxygenase in soybean extracts in relation to the coupled destruction of carotenes. Kies *et al.* (1969) reported that carotene oxidase activity associated with partially purified soybean lipoxygenase appeared to be more susceptible to inactivation by heat than lipoxygenase activity. This indicated the involvement of some other factor in the coupled bleaching of carotenoids. These investigations probably explain some of the earlier studies with soybean lipoxygenase preparations in which a lipoperoxidase factor was considered responsible for the concurrent breakdown of linoleate hydroperoxide and the bleaching of carotene (Blain and Barr, 1961; Blain and Styles, 1959). This factor is probably identical with the enzyme hydroperoxide isomerase identified by Zimmerman (1966). Carotene bleaching activity is now known to reside in the lipoxygenase fraction only (Imamura and Shimizu, 1974).

The attractive red color of some species of fish is attributed to the presence of carotenoid pigments in the skin. Unfortunately, the fading of the skin color during processing and handling poses a serious problem with respect to the acceptability of these fish. Tsukuda and Amano (1966) examined these pigments in the skin of 18 species of fish and reported the major carotenoids to be astaxanthin and tunaxanthin. Any discoloration in these fish was attributed to deleterious changes in these pigments. Tsukuda and Amano (1967) proposed that a tissue enzyme was involved in the discoloration process since the addition of cyanide or heat treatment eliminated such color changes. Further studies by these investigators on defatted fish skin tissues indicated the importance of lipids in these reactions. Tsukada and Amano (1968) confirmed this by studying the effect of unsaturated fatty acids on the discoloration of astaxanthin, tunaxanthin, and β-carotene in model systems containing defatted skin extracts. The results are illustrated in Fig. 2.1 and indicate that the enzyme required

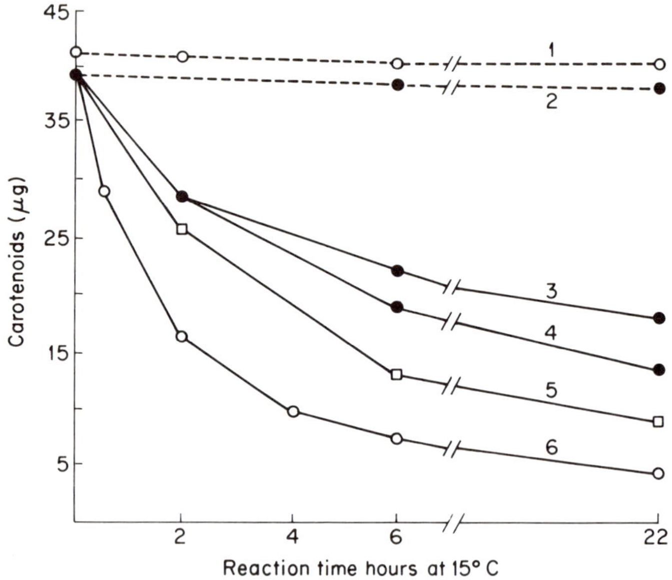

Fig. 2.1. Effect of extracts of defatted rockfish skins with added methyl esters of highly unsaturated fatty acids on the concentration of β-carotene, tunaxanthin, and astaxanthin (Tuskada and Amano, 1968). Curve 1, control (butter solution + tunaxanthin); curve 2, control (β-carotene, heat treated); curve 3, β-carotene + methyl linoleate; curve 4, β-carotene + highly unsaturated fatty acids; curve 5, astaxanthin + highly unsaturated fatty acids; curve 6, tunaxanthin + highly unsaturated fatty acids.

lipids to function normally, which is characteristic of lipoxygenase. The presence of this enzyme in animal tissue has been the center of controversy since carotenoid discoloration can also be brought about nonenzymatically by the action of heme compounds (Bishov *et al.*, 1960; Younathan and Watts, 1959). Nevertheless, evidence has been presented for the location of this enzyme in both the dark muscle of herring and the liver of mackerel and sea bass (Khan, 1952; Ono and Nagayama, 1957a,b, 1959; Shimizu and Fukuhara, 1957a,b).

B. Nonenzymatic

Carotenoids, as mentioned earlier, are very susceptible to oxidation with the concurrent loss of color. Among the products formed are hydroperoxides and carbonyl compounds in addition to other malodorous and volatile components. These pigments are fairly stable under processing conditions (for example, in the case of frozen or heat-sterilized foods), but dehydrated fruits and vegetables tend to exhibit extremely poor carotenoid stability. Mackinney *et al.* (1958) reported a 21% loss of carotenoids when powdered, dehydrated carrots were stored in air for a month compared to zero loss when stored under vacuum.

Evidence in recent years has indicated that mere dehydration of a food product does not eliminate all deleterious changes. Salwin (1959) demonstrated that a minimal moisture level was required by most food products to ensure optimal storage stability. This level was reported by later workers to prevent lipid oxidation from occurring (Koch, 1962; Maloney *et al.*, 1966). A direct relationship between water activity (a_w) and carotenoid loss was subsequently proposed by Martinez and Labuza (1968) for freeze-dried salmon. These investigators found that water acted as a protective agent, preventing the oxidation of the lipid fraction. Carotenoid destruction under these conditions was nonenzymatic, so that the influence of a_w on carotenoid stability became a critical factor.

Chou and Breene (1972) carried out a series of experiments on the oxidative decoloration of β-carotene in low-moisture systems in an attempt to control these reactions. The extent of decoloration was determined by measuring the changes in spectral reflectance of avicel–carotene model systems. Since the reflectance curves for both the oxidized and unoxidized systems were identical in shape, the effect of oxidation appeared to be the same as the effect of decreasing the concentration of β-carotene. These workers followed the reflectance changes in these model systems over a period of 10–20 days of storage at 20°C and at two levels of a_w. The addition of antioxidant to control any oxidative changes was also investigated. Following a short induction period the rate of β carotene decoloration appeared to follow a first-order reaction for all the systems studied. The sorbed moisture and antioxidant in these systems extended the induction period and reduced the rate of oxidative decoloration in a manner similar to that observed for lipid oxidation (Fig. 2.2).

The loss of extractable pigments during the storage of dehydrated ground chili peppers *(Capsicum annum)* was investigated by Chen and Gutmanis (1968). The major carotenoids present, capsanthin, capsorubin, β-carotene, zeaxanthin, and cryptoxanthin, were extracted by the use of organic solvents and determined quantitatively. The loss of color observed during storage at 40°C in the dark over a 12-month period and at 4.0–4.7% moisture appeared to result from oxidative reactions. Color deterioration of the dehydrated chili peppers was not a simple first-order reaction and appeared to be far more rapid in an oxygen-containing atmo-

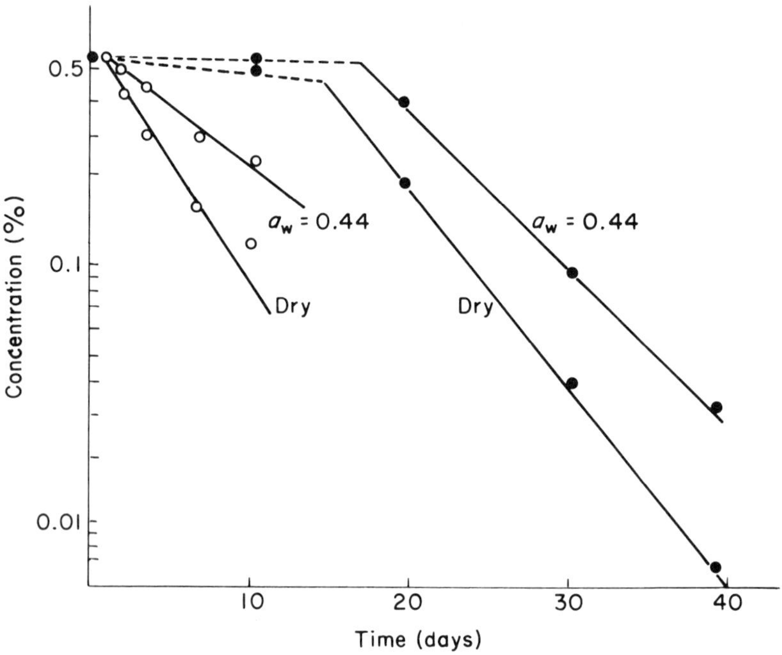

Fig. 2.2 Decoloration of β-carotene in model systems. Dry refers to systems held over solid $CaCl_2$; $a_w = 0.44$ refers to systems held over saturated K_2CO_3 (○; control; ●; BHT—Butylated Hydroxytoluene). (Chou and Breene, 1972. Copyright © by Institute of Food Technologists.)

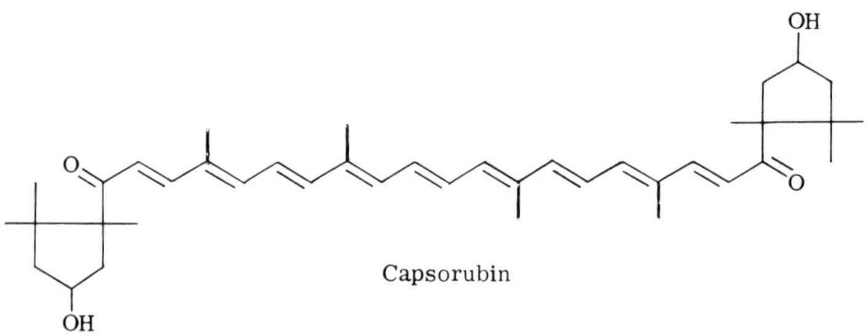

Capsorubin

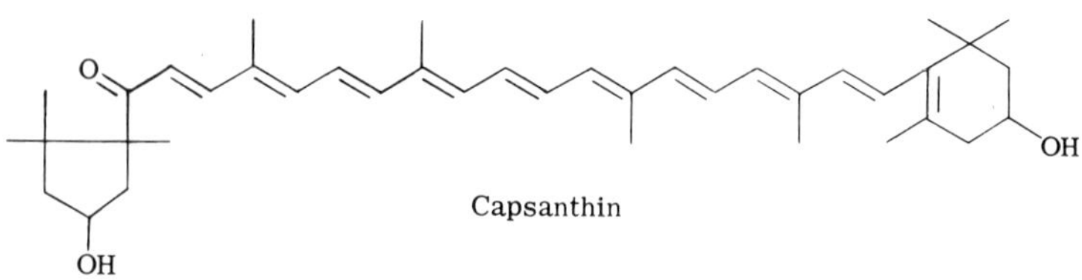

Capsanthin

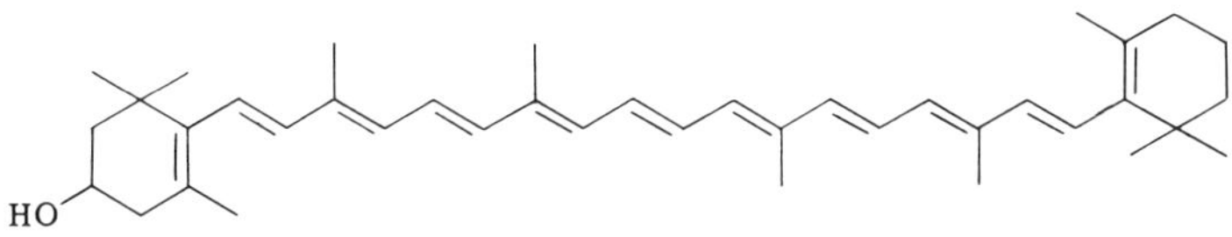

Cryptoxanthin

sphere. When these samples were autoclaved and then stored in air, they proved to be just as unstable as the control samples, indicating spontaneous oxidative degradation. These workers also attempted to control these reactions by the addition of antioxidants, and they reported ethoxyquin and diphenyl-*p*-phenylenediamine to be the most effective. A marked improvement in the surface color of paprika samples treated with ethoxyquin was also observed, as determined with a Hunter Color Difference Meter. As the surface became lighter, there was a corresponding decrease in redness and an increase in yellowness. The ratio of redness/yellowness remained much higher in samples treated with the antioxidant. The use of ethoxyquin at the maximal level of 100 ppm permitted by the U.S. Food and Drug Administration (Anonymous, 1965) appeared to afford good protection against oxidative reactions.

A problem peculiar to the spice industry is the loss of color observed when paprika is exposed to sunlight. This bleaching effect was studied by De La Mar and Francis (1969). Commercial samples of paprika in aluminum-lined trays covered with Pyrex glass were exposed to sunlight over a period of 3 months and then stored in airtight, clear-glass containers. The bleached samples exhibited a greater than 95% loss in total extractable pigments compared with the unbleached samples and were associated with a large number of degradative products. These included isomers of the oxygenated derivatives of capsanthin and capsorubin and were ketonic in nature.

III. ISOMERIZATION

The effect of isomerization reactions on the quality grading of processed pineapples was reported by Singleton and co-workers (1961). They originally observed that the absorption spectrum of extracted pigments from the canned product differed from that of the fresh product (Fig. 2.3). The loss of a peak at 466 nm together with the retention of those at shorter wavelengths made it possible to carry out spectrophotometric determina-

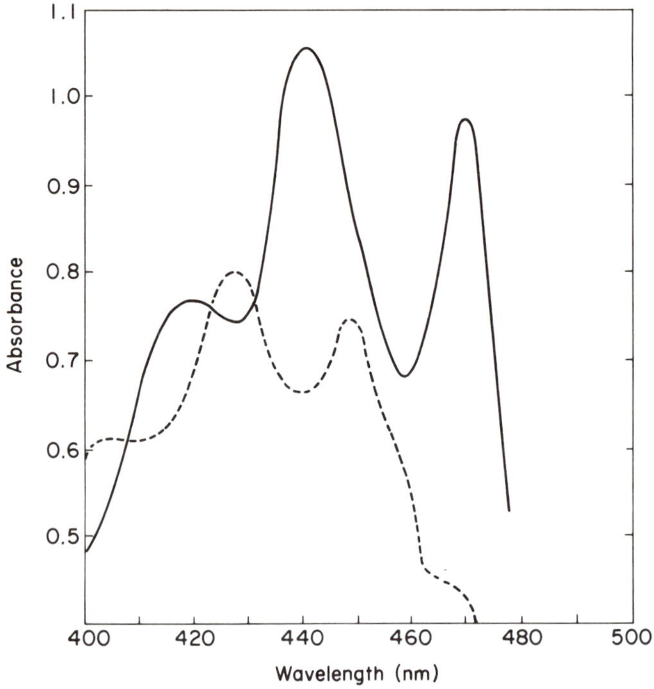

Fig. 2.3. Change in absorption spectrum of pineapple carotenoids during isomerization. Tissue homogenates were allowed to stand at 24°C for various periods before extraction with petroleum ether. Solid line, after 3 minutes; dashed line, after 4 hours and 22 minutes. (Singleton *et al.*, 1961. Copyright © by Institute of Food Technologists.)

tions on the relative amounts of the unisomerized and isomerized pigments present. Although this spectral shift was not obvious to the human eye, it nevertheless resulted in a slight shift of color from an orange-yellow in the fresh pineapple to a more lemon yellow in the canned product. The catalytic effect of acid in these reactions led these investigators to extract the carotenoids from fresh pineapple with hydrochloric acid. This resulted in a partial conversion of the yellow pigments to a blue solid, which was soluble in methanol but insoluble in water or petroleum ether. By treating this blue substance with alcohol the yellow color was restored, although it exhibited a spectrum similar to that of the canned product. This spectral shift was attributed to the isomerization of carotenoids possessing epoxide groups to furanoid rearrangement products. When plant tissue is damaged through handling or processing, the cell vacuole acids presumably come into contact with the carotenoid-bearing plastids, bringing about these reactions.

In the presence of acid the all-*trans* form of the carotenoids isomerizes to the corresponding cis form. The resulting change in shape tends to reduce the resonance in the molecule as well as the intensity of color. This is best illustrated by all-*trans*-β-carotene, which changes from a red-orange to a paler yellow-orange when converted to the corresponding *cis* isomers neo-β-carotene U and neo-β-carotene B. In addition to a change in color other effects, including flavor deterioration as well as a change in absorptive power on chromatographic columns, have been reported (Joyce, 1954; Polgar and Zechmeister, 1942; Zechmeister and Tuzson, 1939). Although autoxidation was recognized as a major factor in color losses associated with these pigments, the effect of cooking was generally considered insignificant. Borchgrevink and Charley (1966), however, suspected that heating carrots in the presence of some of the organic acids might bring about isomerization of some of the red-orange all-*trans* isomers to the less colorful *cis* isomers. These investigators prepared carrots by three different cooking treatments including (a) heating in a covered aluminum saucepan with 150 ml water for 19 minutes, (b) heating in a 2-quart aluminum pressure saucepan containing 120 ml water for 50 seconds at 15 lb pressure, and (c) heating in the same pressure saucepan with 120 ml water for 2 minutes.

The carrots cooked in the covered saucepan were found to have the highest levels of crude carotene and all-*trans*-β-carotene compared to the carrots pressure cooked for 2 minutes, which exhibited the lowest amount of crude carotene and the highest amount of neo-β-carotene. The difference in the observed color appeared to be the result of changes in the all-*trans*-β-carotene as determined quantitatively. The overcooked carrots in treatment (c) were associated with the lower concentration of the all-*trans* isomer. This could be attributed only in part to isomerization to the paler *cis* forms, so that the color changes observed appeared to result from both autoxidative and isomerization changes. Klaui (1971) commented, however, that in general the effects of cooking on color and carotene content were insignificant.

The characteristic red color of salmon muscle is due to the carotenoid astaxanthin, which undergoes degradation during processing, resulting in color losses. Kanemitsu and Matsuura (1961) reported a 36% loss of astaxanthin during the canning of red salmon. They isolated three pigment fractions by chromatography on calcium carbonate. Two of these fractions corresponded to the *cis* isomers of astaxanthin, whereas the third repre-

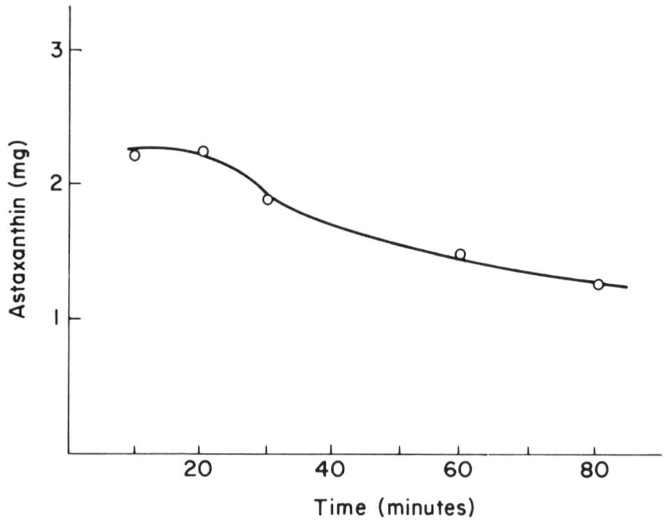

Fig. 2.4. Thermal degradation of astaxanthin during heating at 115°C (Suzuki *et al.*, 1969).

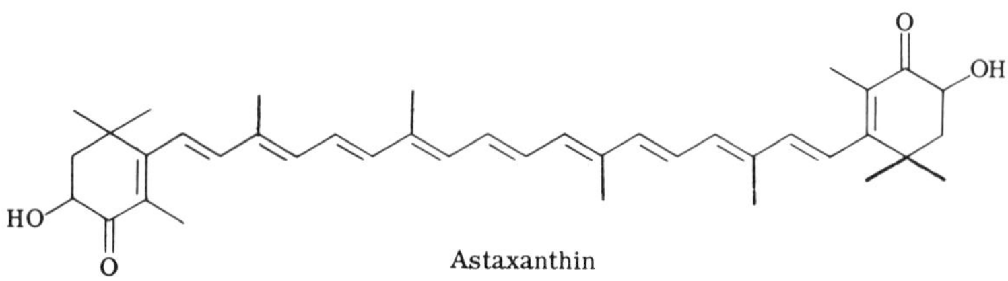

Astaxanthin

sented the unchanged all-*trans*-astaxanthin. Suzuki *et al.* (1969) assessed the color of red salmon by determining changes in astaxanthin content. The level of this pigment was found to decrease rapidly when prepared at 115°C, as illustrated in Fig. 2.4.

IV. PHYSICAL STATE OF CAROTENES

Some vegetables containing high levels of carotenoid were found to exhibit marked color changes when heated in water which could not be explained by either a change in carotenoid content or isomerization to the *cis* isomers (Della Monica and McDowell, 1965; Purcell, 1962). This was observed for carrots and sweet potatoes, which changed from orange to yellow, and for tomatoes, which changed from red to orange-red. Purcell *et al.* (1971) attempted to explain this phenomenon by studying the color shift spectrophotometrically and microscopically. It appeared that the carotene is located in discrete chromoplasts, which when heated disintegrate so that the carotene dissolves in the cellular lipid. The color shift was attributed to the change in the physical state of the carotenoids caused by chromoplast disruption, which resulted in a solution of carotenes in the cellular lipids. In the case of seafoods the change in color is attributed to the disruption of protein–carotenoid complexes. An example is lobster, which changes from a dull olive green to a bright pink when cooked. This color shift is due to the denaturation of the protein in the protein–carotenoid complex and the subsequent release of the bright pink pigment astaxanthin.

Chapter 3

Benzopyran Derivatives: Anthocyanins and Flavonoids

I. INTRODUCTION

Among the benzopyran derivatives are found the water-soluble anthocyanin pigments. These pigments impart the pleasing red and blue colors associated with many fruits and vegetables, which the consumer recognizes as a sign of good quality and freshness. Anthocyanins are polyphenolic substances composed of glycosides of anthocyanidins. The latter are substituted flavylium salts differing in the number and position of free or methylated hydroxyl groups. The structural formula of the flavylium cation showing the ring-numbering convention is illustrated below. Anthocyanidins or aglycones are far less stable than anthocyanins

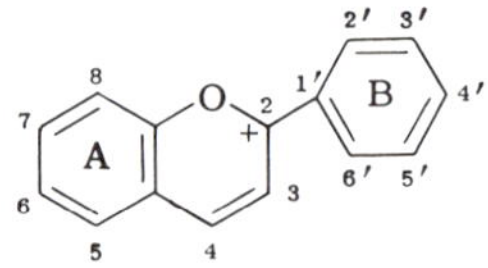

Flavylium cation

and exist as glycosides in the plant tissue in which a sugar moiety is generally linked through position 3 or positions 3 and 5 in the ring. In the majority of foods investigated six anthocyanidins or aglycones occur, as shown in Scheme 3.1.

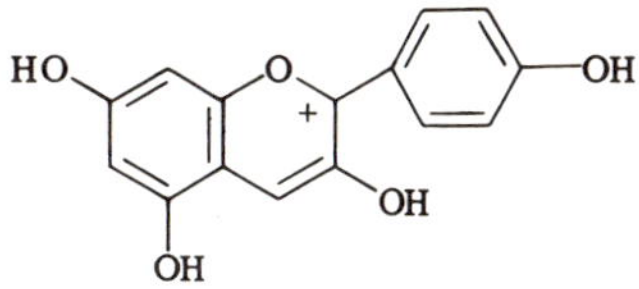

Pelargonidin (3, 5, 7, 4′-tetrahydroxyflavylium cation)

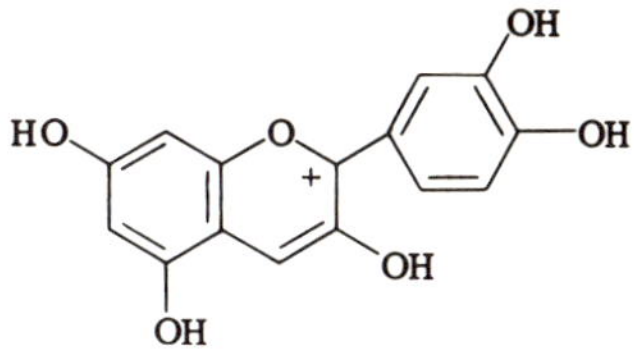

Cyanidin (3, 5, 7, 3′, 4′-pentahydroxyflavylium cation)

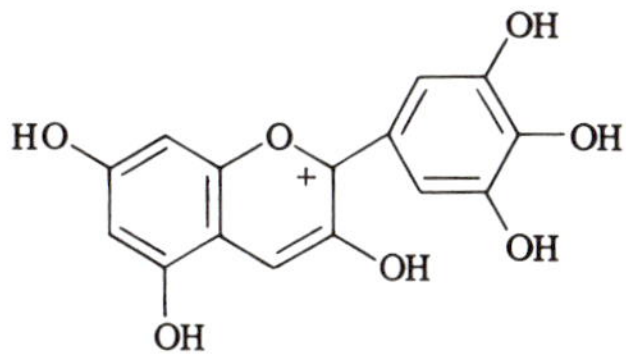

Delphinidin (3, 5, 7, 3′, 4′, 5′-hexahydroxyflavylium cation)

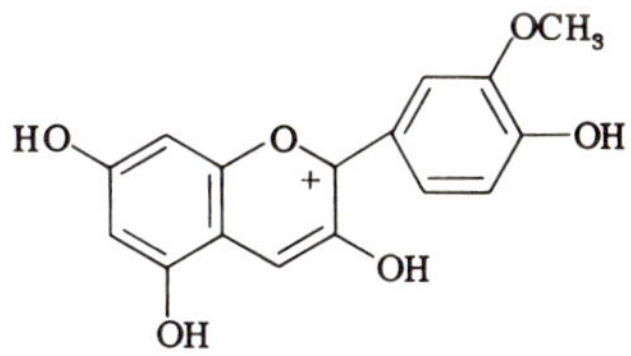

Peonidin (3, 5, 7, 4′-tetrahydroxy-3′-methoxyflavylium cation)

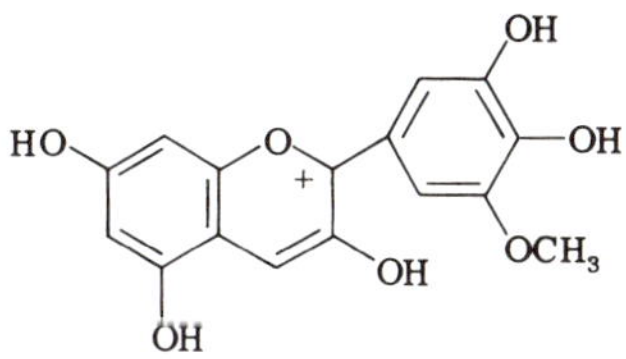

Petunidin (3, 5, 7, 3′, 4′-pentahydroxy-5′-methoxyflavylium cation)

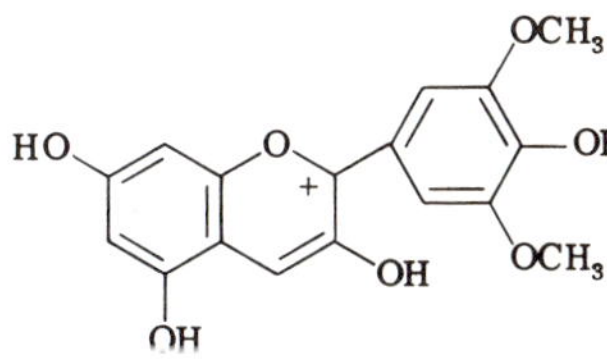

Malvidin (3, 5, 7, 4′-tetrahydroxy-3′, 5′-dimethoxyflavylium cation)

Scheme 3.1. Major food anthocyanidins.

Anthocyanins exhibit maximal stability under acidic conditions, although they are generally unstable and decompose to form insoluble brown-colored products. Such changes can occur under conditions of processing and storage when these pigments are released from the damaged plant tissue and undergo chemical decomposition. Several factors appear to influence the stability of anthocyanins and are discussed in this chapter. An understanding of such changes is mandatory for the food technologist in order to minimize discoloration during the course of processing. These changes can be mediated both enzymatically and chemically.

II. ENZYMATIC DEGRADATION

Several enzyme systems have been shown to decolorize the anthocyanin pigments of fruit tissue. Peng and Markakis (1963) studied the effect of phenolase on red tart cherry anthocyanins. They isolated two pigments from red tart cherry, cyanidin-3-gentiobioside and cyanidin-3-rhamnoglucoside, which they incubated with mushroom phenolase. Although both pigments proved to be poor substrates for the enzyme, in the presence of catechol rapid decolorization occurred. Since catechol is recognized as the normal substrate for phenolase, these authors proposed that the mechanism involved was an indirect one in which the oxidized substrate exerted its effect on the anthocyanin:

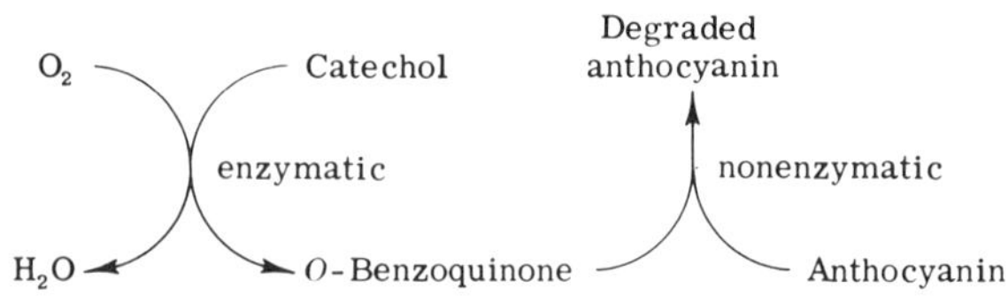

Sakamura *et al.* (1966) demonstrated the oxidation of delphinidin-3-*p*-coumaroylrutinoside-5-glucoside, the major anthocyanin of eggplant, by a polyphenol oxidase preparation isolated from that fruit. Pifferi and Cultrera (1974) monitored the degradation of cyanidin-3-glucoside extracted from Mora var. cherries by two polyphenol oxidases isolated from the pulp of MORBIANE var. clingstone cherries. One of the enzymes showed optimal activity at pH 3.9 in the presence of chlorogenic acid, catechine, and pyrocatechol, with the order of reactivity being chlorogenic acid > catechine > pyrocatechol. The extent of pigment degradation in the presence of these phenolic substrates followed the identical order. The maximal effect observed for each phenol, however, occurred at a pH considerably higher than that required for maximal polyphenol oxidase acivity. The second polyphenol oxidase preparation exhibited a much higher pH optimum: pH 7.0 in the presence of pyrocatechol and pH 7.5 with D(+)-catechine and chlorogenic acid. In this case the pH required for maximal anthocyanin degradation was lower than that required for optimal polyphenol oxidase activity. The pigment degradation/enzyme activity ratio indicated that the anthocyanin underwent extensive breakdown over the range pH 4.5–7.0, as was evident for the other polyphenol oxidase system. Since anthocyanin destruction appeared to parallel the enzymatic oxidation of chlorogenic acid, these workers concluded that the degradation of anthocyanin followed a mechanism similar to that proposed by Peng and Markarkis (1963) and Sakamura *et al.*, (1965).

The decolorization of anthocyanins by glucosidases (anthocyanase) was reported by Huang (1955, 1956). An enzyme isolated from *Aspergillus niger* was found to exert a decolorizing effect on pigment extracts obtained from berry juice as well as on the isolated anthocyanin pigment of blackberry, cyanidin-3-monoglucoside. This enzyme was subsequently referred to as anthocyanase and catalyzed the hydrolysis of the anthocyanin to aglycone and sugar moieties, as illustrated for cyanidin-3-monoglucoside. Since it is labile, the aglycone is rapidly decomposed to colorless products.

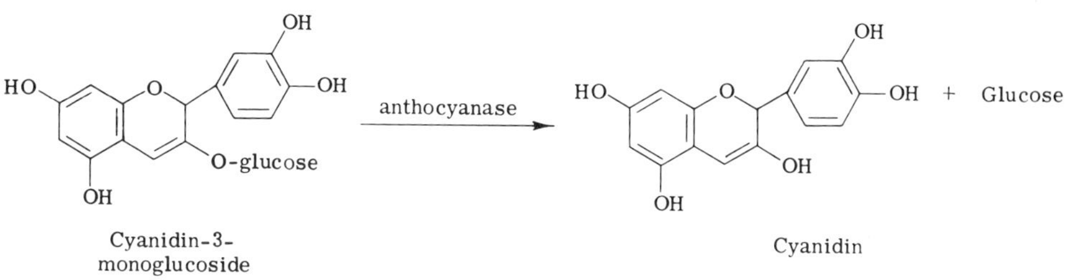

Huang (1955) suggested that this enzyme might be useful in producing white wines from mature red grapes in areas where white varieties were too difficult to cultivate. Yang and Steele (1958) investigated the removal of excessive anthocyanin pigments using enzyme preparations on such concentrated products as blackberry jams and jellies, where pigmentation was too dark and unattractive. They found that treatment of these products with anthocyanase produced brighter, redder, and far more appealing jams and jellies. Wagenknecht and co-workers (1960) later suggested that anthocyanase was in part responsible for the loss of red color associated with scald in sour cherries.

Peroxidases have also been connected with the decoloration of anthocyanins. Nagai (1921) reported that aqueous extracts of scarlet poppies were decolorized by a crude enzyme preparation obtained from the hypocotyls of soybean seedlings in the presence of hydrogen peroxide. This enzyme is located in many plant tissues and has been found in red tart cherries. Grommeck and Markakis (1964) investigated the effect of peroxidase on the stability of anthocyanins isolated from red tart cherries and strawberries. The purified anthocyanins were characterized as cyanidin-3-gentiobioside, cyanidin-3-rhamnoglucoside, and pelargonidin-3-glucoside and incubated with horseradish peroxidase to determine the decolorizing effect. Significant color losses were observed throughout, with a maximum over a pH range of 4.5–5.5, a temperature range of 60°–70°C, and a hydrogen peroxide concentration range of 10^{-4}–10^{-3} *M*.

Since the enzymes capable of decolorizing anthocyanins are present in many fruits and vegetables, the possibility of inactivating them and thereby stabilizing these pigments was investigated. Siegel and co-workers (1971) attempted to stabilize the anthocyanins responsible for the attractive red color of Montmorency tart cherries. These cherries are generally frozen before processing and when required are removed and processed, the latter involving heating to high temperatures to destroy the enzymes. An even more common practice, however, appears to involve merely a slow thawing procedure in addition to comminution of the unheated fruit for infant food. The result is an inevitable loss of color in the fruit. These workers blanched the fruit before freezing in an attempt to minimize the postfreezing color losses normally observed with these fruits. Three blanching periods of 30, 45, and 60 seconds in 100°C steam were carried out and the process repeated in the presence of oxygen, air, and nitrogen. Siegel and co-workers concluded that a blanching period of

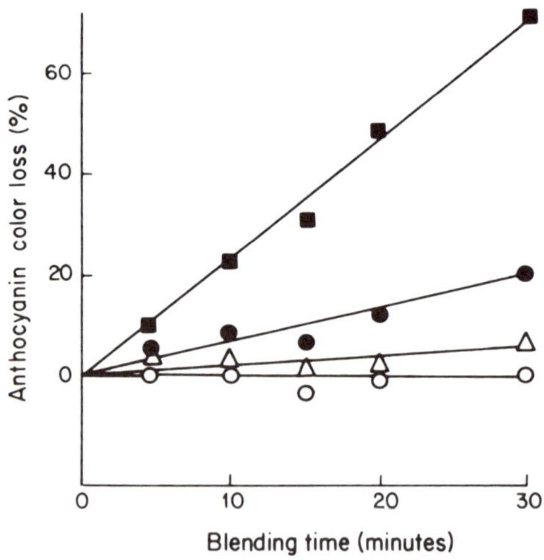

Fig. 3.1. Anthocyanin loss in red tart cherries. Cherries were steam-blanched for 0 (■), 30 (●), 45 (△), and 60 (○) seconds, frozen, and then thawed in a Waring Blendor under oxygen. (Siegel *et al.*, 1971. Copyright © by Institute of Food Technologists.)

45 or 60 seconds was adequate to prevent any anthocyanin decoloration during a 4-hour thawing period at room temperature (Fig. 3.1). Color losses were considerably smaller when blending was carried out under nitrogen than were those when blending was performed under air or oxygen.

III. NONENZYMATIC DEGRADATION

The loss of color associated with processed fruits and vegetables is the result of chemical degradative reactions. Anthocyanins are generally unstable, and their natural red or blue colors can change to an undesirable brown. Such changes are determined by many factors including pH, temperature, ascorbic acid, sugars, oxygen, and metallic ions.

A. Effect of pH

These pigments behave as cations in that they are adsorbed by cation-exchange resins and migrate toward the negatively charged electrode in electrophoresis experiments. They are ulso undeniably phenols and form complexes with certain cations, which is a typical property of phenols. Under acidic conditions, pH 2.0 and above, the colors of anthocyanin solutions are red, and these change from the orange-red of pelargonidin to more purple-red colors with an increase in hydroxylation of the B ring of the molecule. As the pH of the anthocyanin solutions increases, the colors tend to change, usually to blue or purple, although sometimes yellow colors are obtained. Harper (1968) suggested the equlibria shown in Scheme 3.2 to account for the observed changes in color for pelargonidin chloride.

Over the pH range 1–3 the pigment exists as a stable, deeply red oxonium ion (**I**), which is hydrated at a higher pH, forming a colorless pseudobase (**II**). The latter structure is in equilibrium with its keto form (**III**) and can undergo ring opening to form the α-diketone **IV**, which is present in solution over the pH range 3–7. The anhydro base **V** is found in appreciable quantities above pH 7 and is responsible for the purple color. This color fades below pH 7 as a result of the formation of the pseudobase **II** and α-diketone **IV**. On raising the pH this purple coloration changes to mauve and finally to blue at pH 10, increasing in intensity by the formation of the ionized anhydro base **VI**. The anhydro base appears to be extremely stable under alkaline conditions, and no ring opening was de-

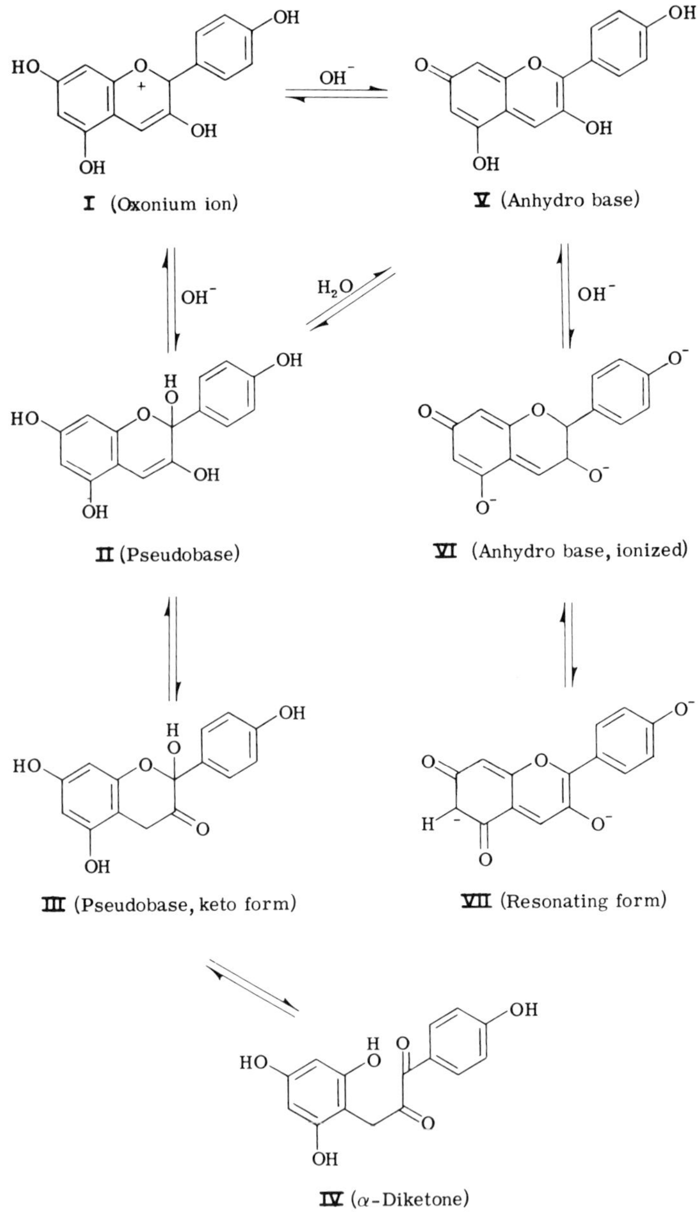

Scheme 3.2. Changes in the molecular structure of pelargonidin chloride with pH (Harper, 1968).

tected up to pH 12.8. Degradation of the ketone, however, leads to the production of a brown precipitate. Such information provides an understanding of the color changes that develop among different fruits under various processing conditions.

Tinsley and Bockian (1960) reported that the degradation of anthocyanins was dependent on the amount of the pigment existing in the pseudobase form. Evidence had been presented earlier by Sondheimer (1953) for the existence of an equilibrium between the red flavylium cation (R^+) and the hydrated pseudobase (ROH):

$$R^+ + 2H_2O \longrightarrow ROH + H_3O^+$$

This relationship was further investigated by Lukton and co-workers (1956) with strawberry anthocyanins. They confirmed that the breakdown of anthocyanin in the presence of oxygen was dependent on pH and directly related to the amount of pigment in the pseudobase form but

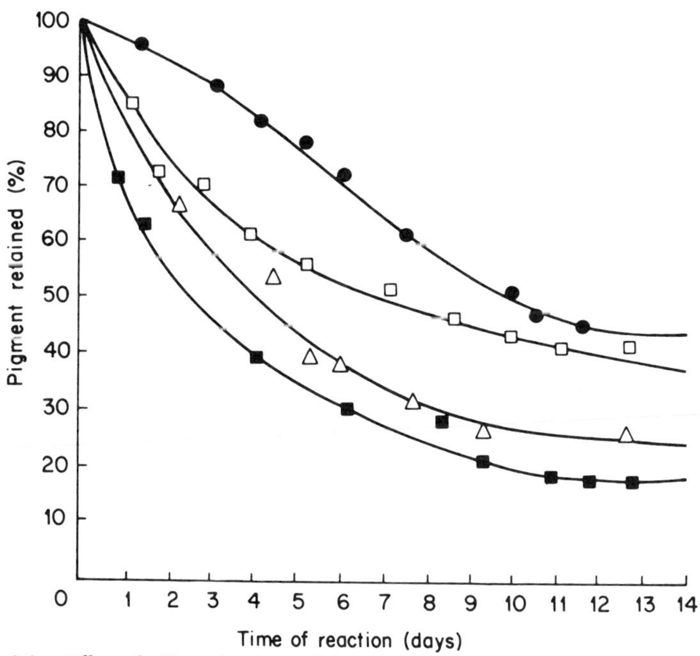

Fig. 3.2. Effect of pH on the thermal destruction of cyanidin-3-diglucoside (in citrate buffer, 0.1 *M*, under oxygen at 50°C). Key: ●, pH 0.95; □, pH 2.15; △, pH 3.25; ■, pH 4.15. (Daravingas and Cain, 1968. Copyright © by Institute of Food Technologists.)

inversely proportional to the amount of flavylium cation. Meschter (1953) had reported previously that the rate of anthocyanin breakdown was dependent on pH and that lowering the pH resulted in a greater degree of stabilization. In studies on raspberry anthocyanins, Daravingas and Cain (1968) similarly observed that lowering the pH significantly retarded the breakdown of the major anthocyanin cyanidin-3-diglucoside (Fig. 3.2). The greater stability of the raspberry anthocyanins in model systems compared to that of the anthocyanins found in the natural juice indicated the involvement of other factors in the destruction of this pigment. Sistrunk and Cash (1968) attempted to improve the stability of anthocyanins in strawberry purée and reported that lowering the pH contributed far more to color stability than did any other method. In an investigation of the color quality of frozen strawberries, Wrolstad and co-workers (1970) found that pH was the only objective measurement that had a positive correlation with the overall quality of the fruit. They further stated that pH was a far better measure of the acid component than was total acidity.

B. Effect of Ascorbic Acid

The contribution of ascorbic acid to anthocyanin degradation in fruit juices was investigated by several researchers. Beattie *et al.* (1943) in studies on processed fruit juices reported a marked decrease in the level of ascorbic acid in the presence of air, which was accompanied by a change in color of the product. They suggested that this resulted from an interaction between the ascorbic acid and anthocyanin pigments present in the fruit juice. Further studies by Sonheimer and Kertesz (1953) indicated that the maximal loss of anthocyanins occurred under the conditions favorable to ascorbic acid oxidation, thereby implicating the importance of the oxidized products rather than ascorbic acid itself. The formation of hydrogen peroxide from ascorbic acid oxidation has been demonstrated by a number of workers and led Sondheimer and Kertesz (1952) to study its effect on strawberry anthocyanins. These workers reported significant color losses, although the formation of hydrogen proxide in strawberry juice was not demonstrated. Pratt *et al.* (1954) also confirmed the participation of ascorbic acid in anthocyanin degradation and implicated riboflavin in these reactions as well. The effect of ascorbic acid on the stability of the strawberry anthocyanin pelargonidin-3-monoglucoside was later investigated by Markakis and co-workers (1957) in the presence and absence of oxygen. Their results are illustrated in Fig. 3.3, in which the extent of pigment degradation in the presence of ascorbic acid and oxy-

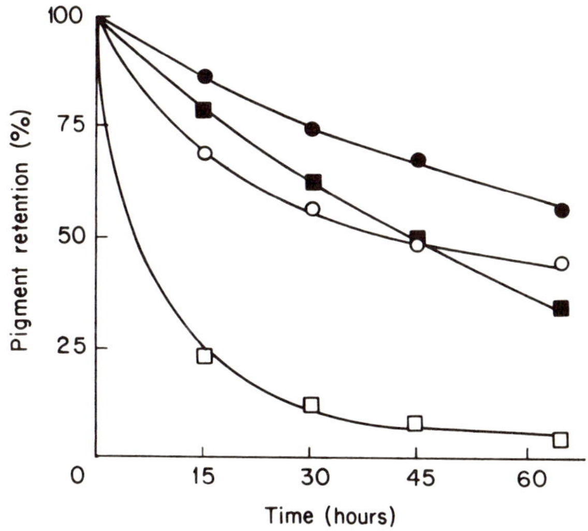

Fig. 3.3. Retention of pelargonidin-3-monoglucoside in citrate buffer (pH 3.40) at 60°C. Key: ●, pigment; ■, pigment + ascorbic acid; ○, pigment + oxygen; □, pigment + ascorbic acid + oxygen. (Markakis *et al.*, 1957. Copyright © by Institute of Food Technologists.)

gen is shown to be greater than in the presence of either oxygen or ascorbic acid. Sistrunk and Cash (1968) attempted to stabilize the color of strawberry purée and reported that the presence of ascorbic acid improved the color retention during the first 6 hours of storage at 50°C. However, a considerable loss of color was evident during subsequent storage for 18 hours. Several chemicals were added to improve the stability of the color including $AlCl_3$, $SnCl_4$, and EDTA. Of these, $SnCl_4$ proved to be the most effective over the pH range 3.0–3.8, although its effect was limited over a longer storage period.

The stability of four anthocyanins of cranberry juice was investigated by Starr and Francis (1968), particularly in relation to the effects of oxygen and ascorbic acid. Cranberry cocktail was prepared according to normal manufacturing procedures using three levels of oxygen headspace and three concentrations of ascorbic acid in the bottles. Pigment losses were found to be higher in the presence of increasing levels of oxygen or concentrations of ascorbic acid. The maximal losses recorded corresponded to the maximal levels or concentrations of oxygen and ascorbic acid used (Fig. 3.4 and 3.5). Although the total pigments were found to decrease sharply on storage, the actual percentage of peonidin and cyanidin galac-

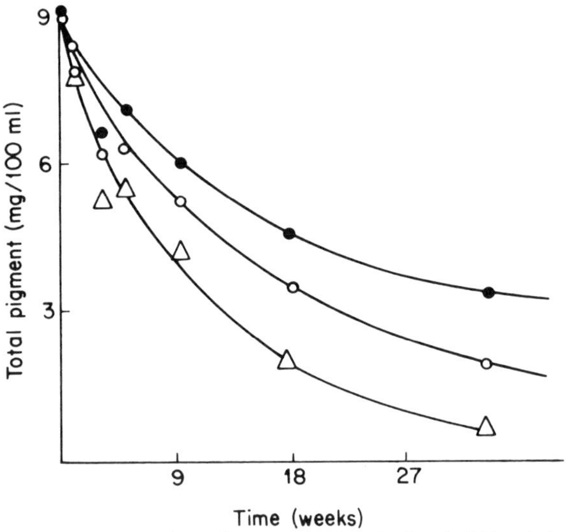

Fig. 3.4. Pigment content of cranberry cocktail packed with 177 μg/ml of ascorbic acid. The different data points represent increasing levels of initial headspace oxygen. (Starr and Francis, 1968. Copyright © by Institute of Food Technologists.)

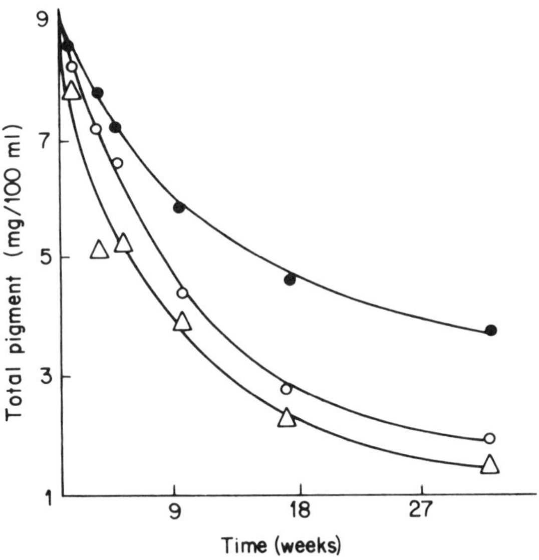

Fig. 3.5. Pigment content of cranberry cocktail packed with headspace oxygen adjusted to 2.0 ml/472-ml bottle. The different points represent increasing levels of ascorbic acid. (Starr and Francis, 1968. Copyright © by Institute of Food Technologists.)

tosides was found to increase, in contrast to peonidin and cyanidin arabinosides, which markedly decreased during these studies.

Wrolstad *et al.* (1970) investigated the effect of several factors, including ascorbic acid, on the color quality of frozen strawberries. Under the low-temperature conditions of their study, they were unable to find any significant correlations between ascorbic acid and color quality. The previously reported accelerated destruction of anthocyanins by ascorbic acid at high temperatures was generally attributed to the corresponding degradation products. These workers therefore suggested that at lower temperatures ascorbic acid may in fact exert a beneficial effect, as demonstrated previously by Sistrunk and Cash (1968).

C. Effect of Sugars

The influence of sugars on the stability of anthocyanins has been investigated. Several sugars were incubated with strawberry pigment at 38°C and the rate of color loss observed. Arabinose, levulose, and sorbose appeared to have a far greater degradative effect on the pigments than either maltose or sorbitol. Tinsley and Bockian (1960) in studies using model systems found that increasing the sugar concentration decreased the amount of extractable strawberry anthocyanins. This loss was attributed in part to the breakdown products, furfural, 5-hydroxymethylfurfural, levulinic acid, and formic acid. Daravingas and Cain (1968) investigated the effects of processing and storage on the stability of raspberry pigments. Significant color changes were observed for several of the pigments, although some of the fractions appeared to be more susceptible than others. Furfural and 5-hydroxymethylfurfural, in addition to being sugar degradative products, are also formed in the Maillard reaction and by the oxidation of ascorbic acid (Eskin *et al.*, 1971). These products appear to condense with anthocyanins to form brown-colored products. 5-Hydroxymethyl furfural may also further decompose to levulinic and formic acids in the acidic environment of most anthocyanin-containing fruits. The effect of glucose and its degradation products on the stability of cyanidin-3-diglucoside, a major anthocyanin of black raspberry, was reported by Daravingas and Cain (1968). The results of their study are shown in Fig. 3.6. The minimal loss of pigment was found in the presence of glucose, whereas the maximal loss occurred in the presence of furfural. Hydroxymethylfurfural did not produce as large a reduction in pigment as furfural, and this was attributed to its breakdown products, levulinic and formic acids, both of which had a smaller effect.

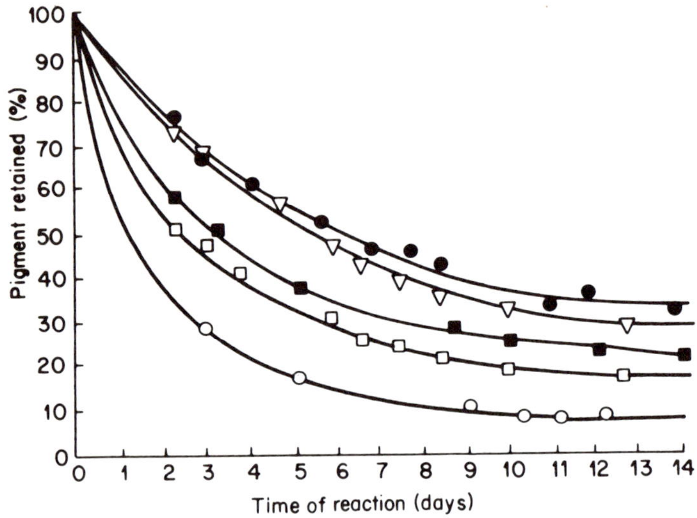

Fig. 3.6. Effect of glucose and certain degradation products on the thermal destruction of cyanidin-3-diglucoside (citrate buffer, 0.1 *M*), pH 3.25, under oxygen at 50°C. In the presence of 0.1 *M* glucose (●), 0.1 *M* formic acid (▽), 0.1 *M* levulinic acid (■), 0.1 *M* 5-hydroxymethylfurfural (□), and 0.1 *M* furfural (○). (Daravingas and Cain, 1968. Copyright © by Institute of Food Technologists.)

IV. METAL COMPLEXING EFFECTS

Color changes in stored food products have been attributed to the formation of metal complexes with anthocyanins or related polyphenolic compounds (Harper, 1957). Anthocyanins containing *o*-dihydroxyl groupings form metal complexes, which results in a shift from red to stable blue and violet pigmentations, as reported for cyanidin-3-glucoside (Jurd and Asen, 1966; Asen *et al.*, 1969; Wrolstad *et al.*, 1970). Can corrosion appears to be a common cause but can be eliminated by coating the tin with lacquer. However, imperfections in the lacquer occasionally result in discoloration problems. Serious cases of discoloration in canned fruits have been reported to involve leucoanthocyanins. These are colorless compounds which when heated in acidic solution are converted to the corresponding anthocyanidins. Small amounts of additional anthocyanidins were evident, however, during the hydrolysis of leucoanthocyanins by alcoholic HCl in studies by Mathews (1969). This was attributed to the formation of simple ethers between anthocyanidins and alcohol of the medium.

Pink discoloration during the commercial canning of processed pears was shown to be due to the decomposition of the leucoanthocyanins present in the fresh fruit to cyanidin (Luh *et al.*, 1960; Nortje, 1964). Chandler and Clegg (1970) investigated this phenomenon in canned pears and

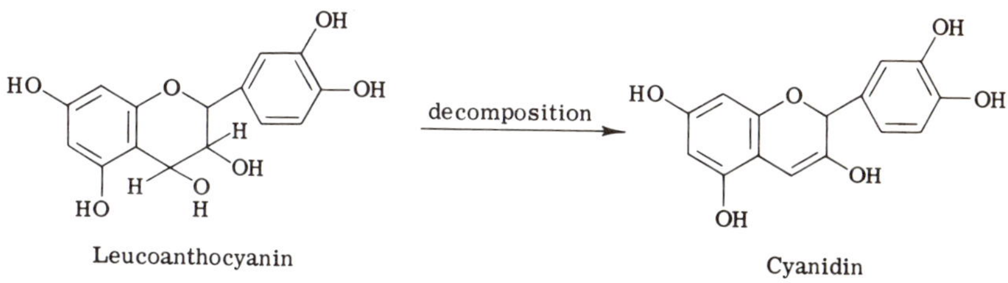

suggested that the pink discoloration was due to the formation of an insoluble tin–cyanidin complex. Van Teeling *et al.* (1971) reported the formation of a pigment complex in processed blueberry syrup. This complex contained glucosides of delphinidin and petunidin in addition to a pectinlike polysaccharide and some proteinaceous material. Aluminum, calcium, iron, and potassium cations, as well as some leucoanthocyanin material, were also involved. This complex exhibited good stability and was reported to be the first tin-free anthocyanin complex isolated from canned fruit.

In addition to anthocyanins and leucoanthocyanins, there are other flavonoid pigments, including anthoxanthins. These colorless and pale yellow pigments are usually present in the cell sap. They differ from anthocyanins by the aglycone moiety consisting of a benzopyrone nucleus. They may be the sole flavonoids in such vegetables as potatoes and yellow-skinned onions or copigments with anthocyanins in red cabbage. Although anthoxanthins can be converted chemically to anthocyanins, the interconversion of these pigments biologically has not been demonstrated. When heated in acidic solution, anthoxanthins yield the corresponding flavone plus one or two molecules of monosaccharides. One of the widely occurring anthoxanthins is quercetin, an important derivative of which is rutin, the 3-rutinoside of quercetin (2-phenyl-3,5,7,3′,4′-pentahydroxybenzopyrone). A greenish-black discoloration observed in canned asparagus was found to be due to the complexing of rutin with iron (Davis *et al.*, 1961). Lueck (1970) reported that the iron derived from the can must undergo oxidation to the ferric state before the discoloration occurs. The formation of the water-insoluble iron–rutin complex appeared to

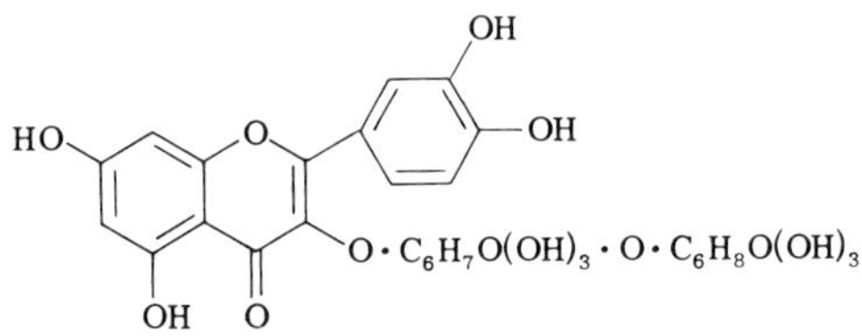

Rutin

be dependent on the relative amounts of tin and iron derived from the can since both appeared to be in direct competition for the rutin molecule. The tin reacted with the rutin molecule to form a bright yellow water-soluble complex, which was still fairly acceptable in appearance. This complex proved to be far more stable than the corresponding iron–rutin complex, so that the addition of small amounts of tin minimized the formation of the undesirable iron–flavanol complex. Lueck (1970) further reported that EDTA added to the brine used in canning asparagus also prevented discoloration by chelating with any metal present.

SECTION II

Natural Flavor Compounds of Foods: Their Biogenesis and Effects on Quality

The flavor of a food can be defined as the subtle interaction of taste and aroma that imparts a pleasing or displeasing sensory experience to each individual. Although the appearance of the food evokes the initial response, it is the flavor that ultimately determines its acceptance or rejection by the consumer. The inhalation of food volatiles together with the release of both volatile and nonvolatile components during chewing elicits the characteristic flavor profile for each food. Considerable information is available concerning the volatile components of fresh and heat-processed foods, whereas the data on the biogenesis of many flavor compounds remain fragmentary. This section is concerned with the biochemical reactions responsible for the generation of the flavors of a selected group of fruits and vegetables.

Chapter 4

Sulfur Compounds: Biogenesis

I. INTRODUCTION

Sulfur compounds possess powerful and distinctive odors which contribute to both the pleasant and unpleasant character of many foods. Whereas fruits produce relatively few sulfur volatiles, the aroma of vegetables is attributed largely to the presence of a variety of sulfur compounds. This chapter discusses the biogenesis of those sulfur compounds contributing to the flavor of raw vegetables in addition to changes induced by processing.

II. ALLIUMS

The genus *Allium* includes such strongly flavored vegetables as onions, garlic, and chives, which are used extensively to enhance the flavor of other foods. These vegetables provide the classic example of enzymatic genesis of flavor compounds in which the products formed undergo nonenzymatic degradation to more stable volatile components such as sulfides, disulfides, trisulfides, and thiosulfonates. Semmler (1892) first reported the presence of disulfides and polysulfides in onion oil, in which allyl-*n*-propyl disulfide was identified among the major products formed. This compound was considered for many years to be the major flavor component of onions. Carson and Wong (1961), however, were unable to confirm the presence of allyl derivatives in onion, but they did identify methyl and *n*-propyl derivatives of di- and trisulfides among the volatiles formed. The distribution of aliphatic mono- and disulfides among the flavor volatiles of *Allium* species was further investigated by Saghir *et al.* (1964) as a possible chemical technique for botanical classification. Using gas chromatography these workers established a taxonomic relationship between the types of alliums and the nature of the alkyl substituent. Garlic varieties were predominant in allyl sulfides and disulfides, whereas onions and leeks were associated with higher levels of *n*-propyl sulfur compounds. The main sulfur volatiles in chives however, were methyl compounds, which was also the case for rakkyo, a Japanese and Chinese pickle. The presence of other sulfur derivatives in each of these vegetables together with other volatiles probably provides the background flavor. More recent studies by Freeman and Whenham (1975) on the volatile components of onion and garlic confirmed the earlier observations of Saghir and co-workers (1964), although there was some discrepancy with respect to the species containing large amounts of methyl compounds which could not be explained by these authors.

The sulfur volatiles discussed so far are not normally present in the intact vegetable tissue, but appear rapidly following disruption of the vegetable during preparation or chewing. It appears that the flavor precursors and their respective enzymes are compartmentalized within the plant tissue and come into contact only when the cellular system is disrupted. The flavor precursors of alliums have been identified as amino acid sulfoxide derivatives which undergo the following general reaction:

$$R{-}SO{-}CH_2{-}CH(NH_2)COOH \rightarrow S\ (\text{volatile}) + NH_3 + CH_3COCOOH$$

S-Alk(en)yl-L-cysteine sulfoxide

Freeman and Whenham (1975) investigated 27 *Allium* species and classified them according to the amino acid precursors present. Three groups were evident based on the predominance of S-1-propenyl-. S-2-propenyl(allyl)-, or S-methyl-L-cysteine sulfoxide as precursor in *A. cepa* (onion), *A. sativum L.* (garlic), or *A. aflaturense B.* Fedtschenko, respectively. These differences are reflected in the volatile composition discussed earlier and are summarized in Table 4.1.

A. S-2-Propenyl(allyl) Derivatives

Cavallito *et al.* (1944, 1945) investigated antibacterial factors in plant tissues and identified a bacteriostatic agent in garlic: allicin. This compound was subsequently characterized as diallyl thiosulfinate and described as irritating to the skin and having an odor characteristic of garlic.

TABLE 4.1

Gas Chromatographic Analysis of *Allium* Species with a High Proportion of *S*-1-Propenyl-, *S*-2-Propenyl(allyl)-, and *S*-Methyl-L-Cysteine Sulfoxide Precursors[a]

Precursor	Species	Relative proportions of alk(en)yl radicals			
		1-Propyl	Methyl	1-Propenyl	2-Propenyl(allyl)
S-1-Propenyl-L-cysteine sulfoxides	*A. cepa* L. (onion, "Rynsburger")	87.9	6.1	6.0	—
	A. cepa L. ("Elsoms 71/94")	89.4	3.5	7.1	—
	A. cepa L. ("Tunisian Red")	90.9	3.9	5.2	—
S-2-Propenyl(allyl)-L-cysteine sulfoxides	*A. sativum* L. (garlic)	2.0	13.4	—	84.6
	A. tuberosum Rottler (ex Sprengel, Chinese chives)	1.0	39.2	—	59.6
	A. ursinum L. (wild garlic, Ramsons)	4.6	31.5	—	63.9
S-Methyl-L-cysteine sulfoxides	*A. aflatunense* B. Fedtschenko	5.9	94.1	—	—
	A. flavum L.	13.0	73.4	13.6	—
	A. pulchellum Don	22.8	63.5	13.7	—

[a] Adapted from Freeman and Whenham (1975).

Although not found in the whole, undamaged garlic bulb, allicin was readily detected when garlic cloves were finely chopped. These researchers concluded that a thermostable precursor of allicin and an activating enzyme must be located in different parts of the plant. Only when the cells were damaged could allicin be produced by the action of the enzyme on the precursor. Stoll and Seebeck (1948) isolated this precursor, (+)-*S*-allyl-L-cysteine sulfoxide, and referred to it as alliin. An enzyme, alliinase, capable of converting alliin to allicin, pyruvate, and ammonia was subsequently identified by Stoll and Seebeck (1948). This enzyme is also referred to as cysteine sulfoxide lyase, or C–S lyase (EC 4.4.1.4). Under normal room temperature conditions, allicin undergoes nonenzymatic degradation to thiosulfonate and disulfides. Although allicin appears to be responsible for the fragrant odor of freshly chopped garlic, it has not been identified as a component of commercial garlic oils. Oaks *et al.* (1964) and Brodnitz *et al.* (1971) separated the components of garlic oil by gas chromatography and reported that mono-, di-, and trisulfides accounted for 90% of the total peak area. These researchers concluded that allicin degraded rapidly during the course of gas chromatographic analysis as well as at room temperature. The combined enzymatic and nonenzymatic degradation of alliin is outlined in Scheme 4.1. The initial stage is the formation of sulfenic acid catalyzed by alliinase. This compound is extremely unstable and undergoes rearrangement to thiosulfinate or allicin, which undergoes spontaneous decomposition to thiosulfonate and disulfides. The disulfides themselves can then decompose to form a complex mixture of monosulfides and trisulfides.

B. *S*-1-Propenyl Derivatives

The production of onion volatiles involves a system identical to that described for garlic, except that in this case the predominant flavor precursor is *S*-1-propenyl-L-cysteine sulfoxide. Freeman and Whenham (1975) pointed out that one of the major differences between onions and other members of the genus *Allium* is that *S*-1-propenyl-L-cysteine sulfoxide gives rise to thiopropanal *S*-oxide as the initial unstable product of enzyme fission, whereas *S*-2-propenyl(allyl), 1-propyl. and methyl amino acid sulfoxide derivatives form the corresponding thiosulfinates under identical conditions. The presence of the lachrymatory factor and the cyclic amino acid cycloalliin was considered to be unique to the onion. Spare and Virtanen (1963) isolated an extremely reactive amino acid sulfoxide. *S*-1-propenyl-L-cysteine sulfoxide, from onion which appeared to

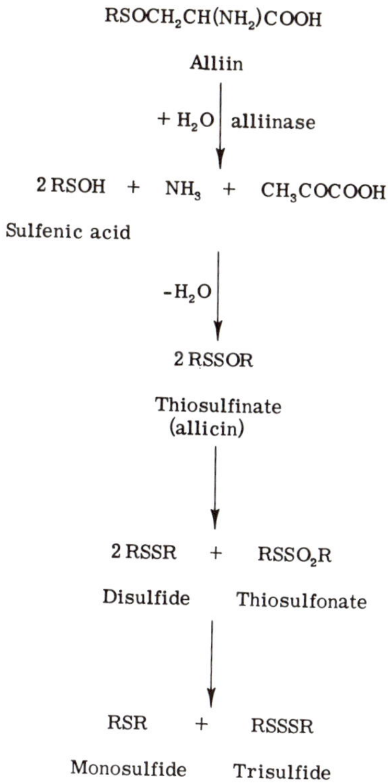

Scheme 4.1. Enzymatic and nonenzymatic degradation of alliin.

be the precursor of the lachrymatory factor. This compound, however, was found to cyclize in the presence of base to form cycloalliin, whereas in the presence of mineral acid the C–S bond disrupted, yielding cysteine, propionaldehyde, and possibly elementary sulfur. It was evident, therefore, that the large amounts of cycloalliin reported by previous investigators were probably produced as artifact from S-1-propenyl-L-cysteine sulfoxide, although trace amounts of this cyclic derivative may be naturally present (MacLeod, 1970). Virtanen and Spare (1961, 1962) demonstrated the formation of the lachrymatory factor together with the production of ammonia and pyruvate when S-1-propenyl-L-cysteine sulfoxide was incubated with onion extract. This reaction appeared to explain both the formation of the lachrymatory factor and the excessive amounts of pyruvate previously unaccounted for.

C. Lachrymatory Factor

The precise nature of the lachrymatory factor eluded scientists for many years, although at one time it was considered to be propenylsulfenic acid. Moisio *et al.* (1962), using mass spectrometry, observed a strong peak of mass 90 after 1 minute of mixing a solution of S-1-propenyl-L-cysteine sulfoxide and onion enzyme. This compound was extremely unstable, and after 1 hour two additional peaks at *m/e* 58 and *m/e* 98 corresponding to propionaldehyde and 2-methyl-2-pentenal were evident. The latter compound was presumed to be formed by aldol condensation of propionaldehyde. Further studies using D_2O indicated that the compound with mass 90 probably corresponded to propenylsulfenic acid (Scheme 4.2, II) with a half-life of 90 seconds. Brodnitz and Pascale (1971) succeeded in isolating the lachrymatory factor by gas chromatography and subsequently identified it as thiopropanal S-oxide by spectral analysis. This structure was confirmed by the chemical synthesis of straight-chain thioalkanal S-oxides, the majority of which possessed lachrymogenic properties. On the basis of this evidence Schwimmer and Friedman (1972) proposed that thiopropanal S-oxide probably results from rearrangement of the sulfenic acid, as shown in Scheme 4.2.

Schwimmer and Friedman (1972) suggested that part of the propyl-containing disulfides predominating in onion distillates could be formed via the lachrymatory factor and not from trace amounts of S-propyl-L-cysteine sulfoxide reported in onions by Carson and Wong (1961) and Matikkala and Virtanen (1967). In addition to being the precursor of the lachrymatory factor, S-1-propenyl-L-cysteine sulfoxide was shown by Schwimmer (1968, 1969) to be predominantly responsible for the sensory

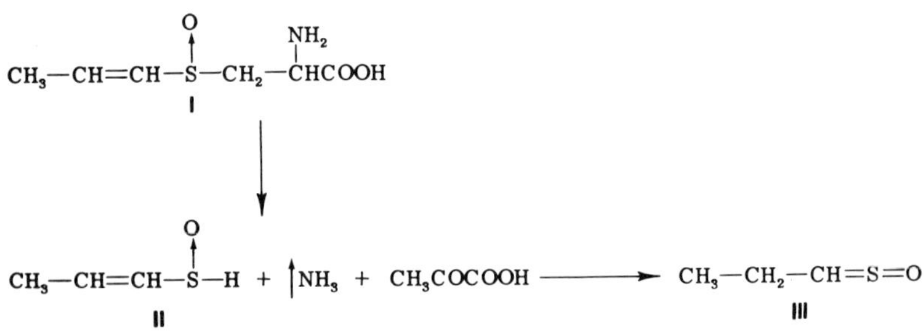

Scheme 4.2. Formation of the lachrymatory factor thiopropanal S-oxide (III). I, Propenyl-L-cysteine sulfoxide; II, propenylsulfenic acid.

attributes of the onion. Incubation of methyl-, propyl-, and S-1-propenyl-L-cysteine sulfoxides with a particular fraction of onion homogenate produced an onionlike odor and burning sensation in the eyes in the presence of propyl and S-1-propenyl derivatives. The lachrymatory effect, however, was evident only in the presence of S-1-propenyl-L-cysteine sulfoxide. Schwimmer (1968) established the importance of the latter compound in onion flavor by monitoring the production of pyruvic acid and bitterness under conditions similar to those existing in the fresh onion homogenate. Preparations of C-S lyase were incubated with S-1-propenyl-L-cysteine sulfoxide as well as in the presence of the corresponding methyl and propyl derivatives at levels equivalent to that reported in onion (Virtanen, 1965; Schwimmer, 1968). The development of odor was apparent during the early stages of the reaction, as was the production of pyruvate, and increased with time. This observation confirmed earlier results indicating a correlation between pyruvate production and odor development in onion systems (Schwimmer and Weston, 1961; Schwimmer and Guadagni, 1962). Bitterness, however, was not evident until the majority of the substrate had been converted to pyruvate, as shown in Fig. 4.1. This indicated the possibility of a secondary reaction in which a second enzyme might be involved in flavor genesis. Kinetic evidence for the role of S-1-propenyl-L-cysteine sulfoxide was provided by Schwimmer (1969) using onion C-S lyase preparations. The affinity of this substrate for the enzyme, as determined by K_m, was almost six times that of the methyl derivative, whereas the V_{max} was four times that of either the methyl or propyl derivative. The relative contribution of the S-1-propenyl sulfoxide to the overall reaction rate in the presence of the other cysteine sulfoxide derivatives was calculated to be 98% on the basis of the quantitative data of Matikkala and Virtanen (1967). The C-S lyases fell into two groups, those in onion exhibiting an optimal pH in the alkaline range and those in garlic being more active at an acidic pH. The inactivation of the onion enzyme under acidic conditions suggested that it was inactivated by some protonated intermediate, possibly still enzyme bound. This protonated derivative was thought to be the reaction product with pyridoxal phosphate, the recognized coenzyme for this enzyme. The mode of action of C-S lyase was proposed by Schwimmer and Friedman (1972), as shown in Scheme 4.3.

Shakaranarayana *et al.* (1974), in a review of the production of sulfur volatiles in foods, referred to a study by Saghir and Mann (1965) on 25 wild species of *Allium*, none of which exhibited any lachrymatory properties. The precursor of the lachrymatory factor was presumed by these researchers to be absent since all of the species possessed considerable

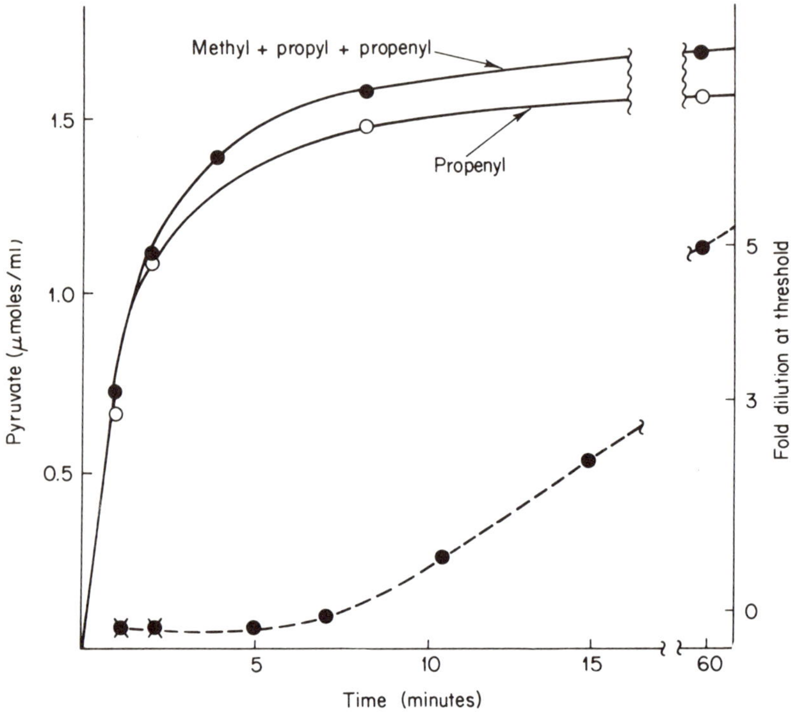

Fig. 4.1. Production of pyruvic acid (—) and bitterness (---) in the presence of S-propenyl-L-cysteine sulfoxide (2.0 μmoles/ml) and the same concentration of the propenyl derivative plus methyl (1.4 μmoles/ml) and propyl derivatives (0.2 μmoles/ml). Bitterness expressed as magnitude-of-fold dilution of reaction mixture at taste threshold. The crosses indicate that no bitterness was detected at zerofold dilution (Schwimmer, 1968).

flavor and aroma, and it was therefore concluded that the lachrymatory factor may not be important to onion flavor. Although this may be true, the presumed absence of the precursor of the lachrymatory factor nevertheless requires further clarification in view of the more recent studies by Schwimmer (1968, 1969).

D. Glutamyl Peptides

The volatiles responsible for onion flavor are formed by the various enzymatic and nonenzymatic reactions discussed previously. Critical to this process is the accessibility of the precursor to enzymatic attack, which determines the overall flavor strength. Matikkala and Virtanen (1967)

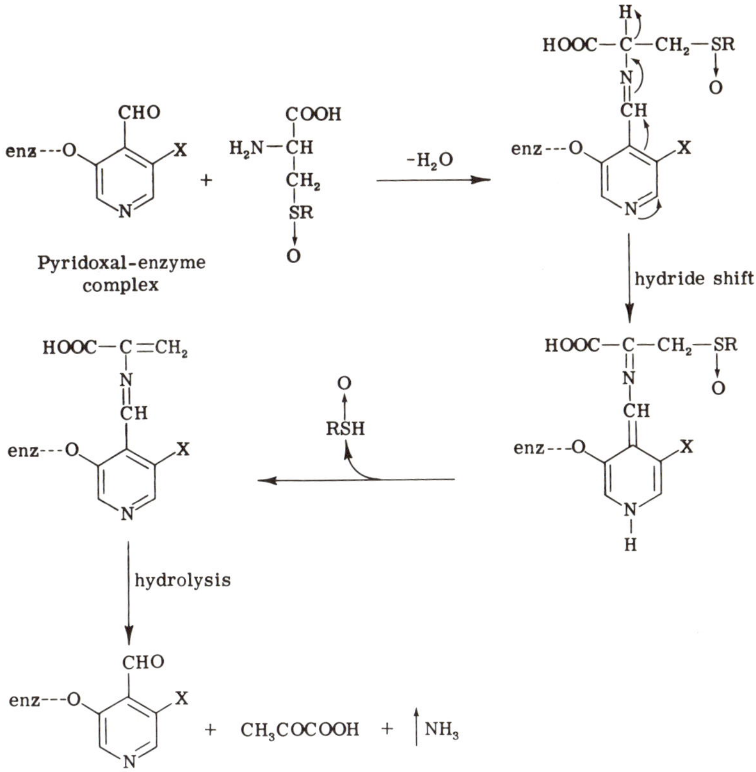

Scheme 4.3. Mode of action of C–S lyase (Schwimmer and Friedman, 1972).

found that in the mature onion almost half of the precursor is bound up as γ-glutamyl peptide, which is not accessible to enzymatic attack. Schwimmer (1971a) reported, however, that the flavor precursor could be released from this peptide by the action of a transpeptidase and then be available for attack by C-S lyase. The ability of these enzymes to release the flavor precursor for further enzyme reaction forms the basis for enhancing and supplementing the lost flavor of dehydrated onion powders as well as other allium powders. This work provides an excellent example of how an understanding of flavor biogenesis can be applied in a practical manner.

The enzyme involved in releasing *S*-alk(en)yl-L-cysteine sulfoxide from the peptide is γ-L-glutamyl transpeptidase (EC 2.3.2.1), which catalyzes the transfer of a glutamyl moiety of the peptide (Scheme 4.4) to a second amino acid. It is widely distributed in mammalian kidney (Orlowski and

Meister, 1965) but is absent from unsprouted onions even though a large number of γ-L-glutamyl peptides are present (Schwimmer and Austin, 1971b). The addition of this enzyme releases the flavor precursor so that it becomes accessible to endogenous C-S lyase activity or a suitable exogenous source, as observed in *Albizza cophanta* seeds by Schwimmer and Guadagni (1968). The coupling of both C-S lyase from *Albizza cophanta* seeds and kidney transpeptidase to generate the flavor was subsequently investigated by Schwimmer (1971a). It did not appear to be particularly promising at first since C-S lyase was optimally active at pH 8.0 in the presence of pyridoxal phosphate and pyrophosphate whereas kidney transpeptidase was optimally active in Tris buffer at pH 9.0 in the presence of Mg^{+}. Schwimmer (1971a), however, showed that at pH 8.0 in the presence of pyridoxal phosphate and pyrophosphate the activity of transpeptidase was equivalent to that at pH 9.0 in the presence of Mg^{2+}. It was now feasible to carry out the two enzyme reactions under the same experimental conditions. Schwimmer (1971a) added these enzymes to dehydrated onion extracts and monitored the increase in pyruvate. In the presence of transpeptidase no pyruvate production was observed, but in combination with C-S lyase there was an increment of 4 μmoles/gm pyruvate produced over and above that produced in the presence of C-S lyase only. This increment, although reflecting the presence of γ-L-glutamyl peptides containing S-1-propenyl-L-cysteine sulfoxide, represented only a small fraction of the glutamyl peptides found in fresh onion before dehydration (Matikkala and Virtanen, 1967). This discrepancy was attributed to several factors including insufficient time for the enzyme to act, varietal differences, peptide destruction during processing, as well as losses incurred during storage. The results of this study (Fig. 4.2) indicate the ability of the appropriate enzyme adjuncts to enhance the full flavor potential of processed onion products.

Schwimmer and Austin (1971a) later identified γ-L-glutamyl transpeptidase activity in sprouted onion that was capable of liberating *p*-nitroanilide from γ-L-glutamyl-*p*-nitroanilide. The enzyme was purified 800-

γ-L-Glutamyl amino acid sulfoxide* + amino acid —transpeptidase→ γ-L-Glutamyl amino acid + amino acid sulfoxide*

Scheme 4.4. Mechanism of γ-L-glutamyl transpeptidase action (asterisk indicates flavor precursor).

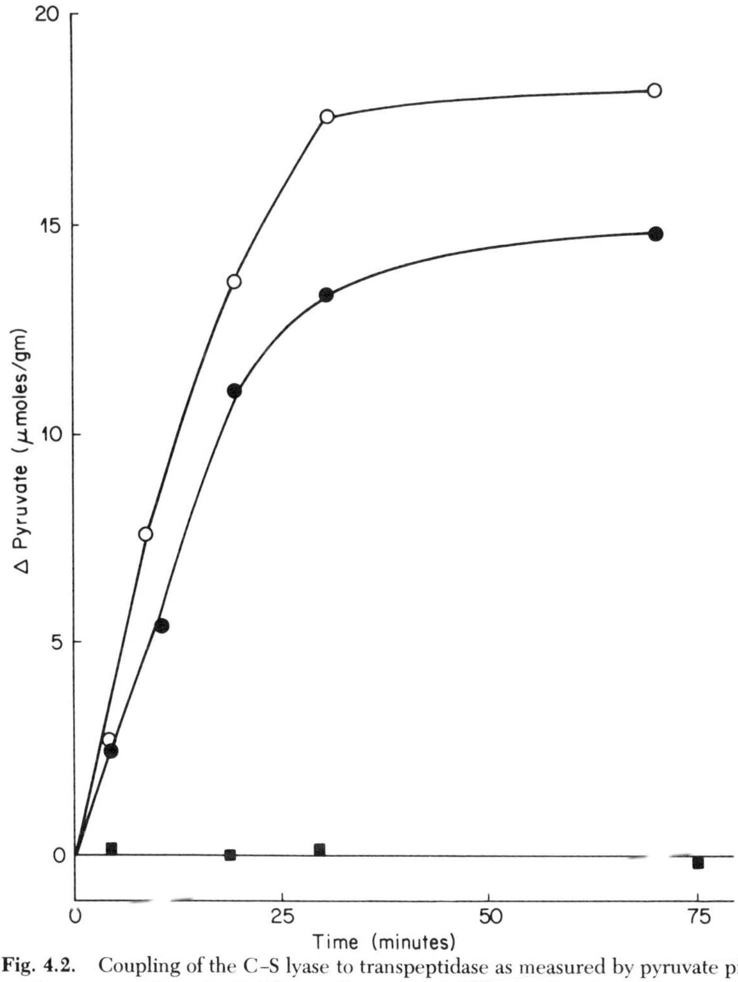

Fig. 4.2. Coupling of the C–S lyase to transpeptidase as measured by pyruvate production (■, transpeptidase; ●, C–S lyase; ○, transpeptidase plus C–S lyase). Reprinted with permission from Schwimmer, *J. Agric. Food Chem.* **19**, 980, 1971. Copyright by the American Chemical Society.

fold and appeared to be a transpeptidase (EC 2.3.2.1), although the crude extract was thought to contain a second enzyme, γ-glutamyl hydrolase, previously found in chive seedlings by Matikkala and Virtanen (1965). This study indicated that, of the nine endogenous γ-glutamyl peptides detected in onion, that containing *S*-1-propenyl-L-cysteine sulfoxide probably disappears in the sprouted onion by way of transpeptidation. A patent was subsequently developed by Schwimmer (1971b) for isolating and utilizing onion γ-glutamyl transpeptidase to enhance the flavor of allium products.

E. *S*-1-Propenyl-L-cysteine Sulfoxide

Evidence for the biosynthesis of *S*-1-propenyl-L-cysteine sulfoxide from the amino acid valine was presented by Granroth (1970). He proposed the mechanism shown in Scheme 4.5.

F. Organogenesis and Onion Flavor Production

The characteristic flavor of onion was found in the root-producing callus isolated by Fridbourg (1971) from ariel bulbs of *Allium cepa* var. *proliferum*. These were grown in agar and liquid cultures on synthetic medium containing a variety of growth regulators. Root formation was stimulated in the presence of naphthaleneacetic acid (NAA), and synergistic effects on the growth and formation of the root of the callus were evident when NAA was treated with indole-3-acetic acid (IAA) or cytokinin. Later, Davey *et al.* (1974) investigated the production of flavor components by the callus. The callus tissue was isolated from root explants of *A. cepa* and propagated in the dark at 25°C. The tissue propagated roots on a medium containing NAA, although no shoot development was evident in any of the growth studies. Examination of flavor production by the undifferentiated callus and tissue with roots showed the pyruvate levels to be markedly lower than in the corresponding fresh onion tissue. Addition of inorganic or organic sulfur-containing compounds to the culture media had no significant effect on pyruvate production. Since the sulfate level in the callus tissue was equivalent to that found in fresh onion, the absence of flavor components could not be attributed to sulfate deficiency (Freeman and Mossadeghi, 1970). These results did not agree with those reported by Friedbourg (1971), but the tissue preparations of Davey *et al.* (1974) differed in their ability to form shoots and their cytology was largely polyploid. This suggested different trends of subculture in these two investigations. Subsequent research by Freeman and associates (1974) was conducted on the flavor components of freshly isolated callus of onion (*A. cepa* L.). The isolate, monitored in culture for 10 months, produced different amounts of flavor components depending on cultural conditions. Only small amounts of flavor components were found in the undifferentiated callus as a consequence of lack of precursor rather than lack of

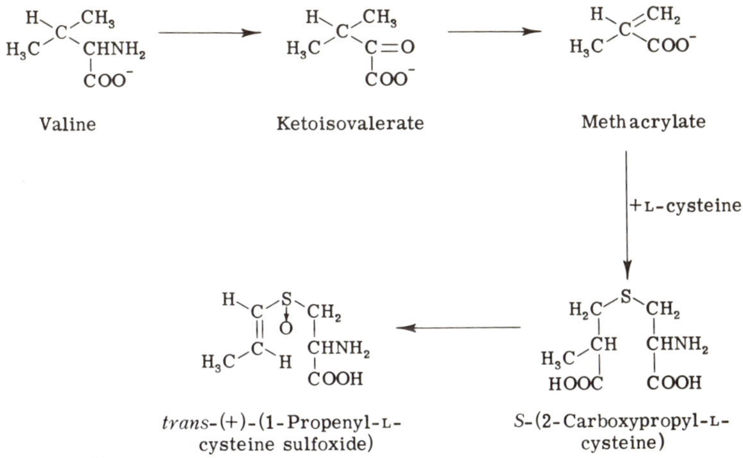

Scheme 4.5. Biosynthesis of *S*-1-propenyl-L-cysteine sulfoxide (Granroth, 1970).

alliinase. The callus tissue with differentiated roots, however, exhibited flavor intensities similar to those of fresh onion, but the lachrymatory potency was low. These results were in agreement with those reported by Fridbourg (1971). The low concentration of flavor precursors in the undifferentiated tissues suggested that root development may be necessary for precursor formation. This is supported by the fact that in alkaloid-producing plants the roots are mainly responsible for alkaloid production (Luckner, 1972). Further research is required to establish the role of the roots in the production of flavor precursors in the onion plant.

C. Water Regime and Flavor Production in Field-Grown Onions

Vegetables grown in areas where there is an abundant water supply are large and lush but lacking in flavor, whereas those grown on restricted amounts of water are smaller and less attractive but richer in flavor components. Freeman and Mossadeghi (1973) conducted field experiments with onions in which the plants were covered by movable, transparent shelters to exclude the rain. Two treatments were applied: (a) restoring the soil to field capacity by overhead irrigation whenever the soil moisture deficit reached 25 mm and (b) not supplying any water (rainfall or irrigation) after the seedlings had been transplanted into soil at field capacity. The bulbs were harvested, and juices expressed from the bulbs were

TABLE 4.2

Effect of Water Regime on the Flavor Strength of Field-Grown Onions[a]

Group	Dry matter (%)	glc headspace, total peak area (cm^2)	Total pyruvate (μmoles/ml)	Taste threshold concentration (ppm)	Refractive Index[b] (n_D^{20})
Water abundant	10.7	100[c]	10.8	8.0	1.34702
Water restricted	12.9	385	13.2	4.9	1.34952

[a] From Freeman and Mossadeghi (1973).
[b] Values ± 0.00006.
[c] Signifies that difference is significant at $P = 0.001$.

subjected to analysis of headspace, total pyruvate production, refractive index, and taste threshold concentrations. As can be seen in Table 4.2 the onions subjected to water stress exhibited a flavor development 3.9 times that of those grown under conditions of water abundance, as determined by headspace total peak areas. This difference is further evident from the taste threshold data. Previous experiments on the effect of water stress had indicated an accumulation of carbohydrates and amino acids in barley (Gauch and Eaton, 1942), turnip leaves (Stewart, 1967), and wilting lucerne leaves (Henrici, 1952). The possibility of low molecular weight metabolites increasing during the period of water stress was suggested by Freeman and Mossadeghi (1973) to explain the increased amounts of flavor precursors. These resulted in increased levels of flavor components released enzymatically when the tissue was disintegrated. Since the flavor precursor is a substituted cysteine, an increase in the level of alkyl-L-cysteine could give rise to the normal flavor sulfoxide components discussed previously.

III. SHIITAKE MUSHROOM C–S LYASE SYSTEM

A novel C–S lyase system has been reported in Shiitake mushroom *(Lentinus edodes)*, a Japanese food delicacy. It appears to be responsible for the formation of lenthionine, the major flavor component identified in this mushroom. The flavor precursor is an *S*-substituted L-cysteine sulfoxide, which, as is the case for *Allium* species, is tied up as a γ-glutamyl peptide. This peptide was isolated by Yasumoto *et al.* (1971a) and re-

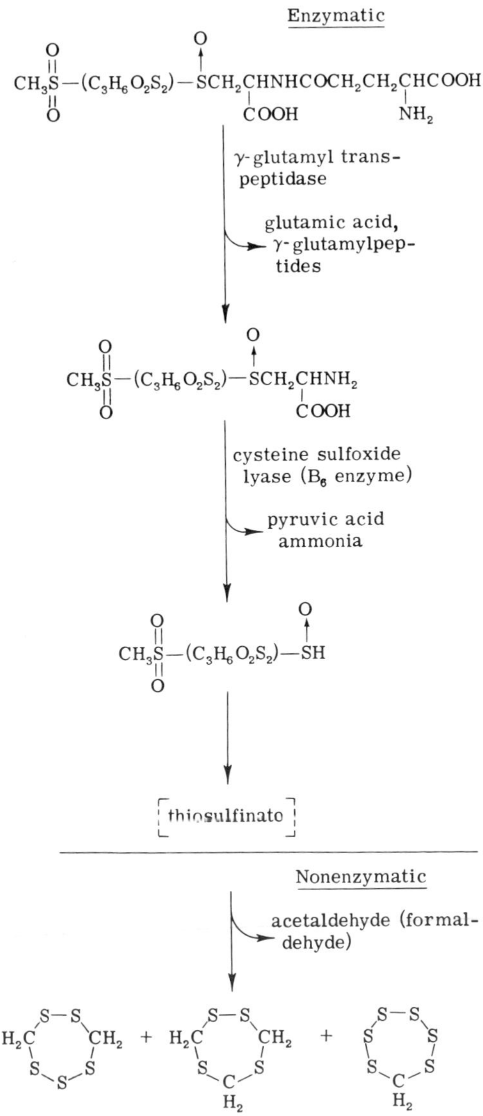

Scheme 4.6. Postulated pathway for the enzymatic and nonenzymatic formation of lenthionine from lentinic acid (Yasumoto *et al.*, 1971b).

ferred to as lentinic acid. The enzymes involved in the conversion of lentinic acid to lenthionine were extracted from fresh fruiting bodies of Shiitake mushroom by Yasomoto *et al.* (1971b). One enzyme involved was a γ-glutamyl transpeptidase, which released the cysteine sulfoxide precursor for further attack by a pyridoxal phosphate-dependent *S*-alkyl-L-cysteine sulfoxide lyase. The latter enzyme catalyzed the formation of an unstable intermediate, thiosulfinate, which underwent nonenzymatic degradation to polythiepanes including lenthionine (Scheme 4.6).

IV. BRASSICA

There has been considerable interest in recent years in characterizing the flavor of vegetables of the brassica family. These include cabbage (*Brassica oleracea* L.). Brussels sprouts (*Brassica oleracea* var. *gemmifera*), cauliflower (*Brassica oleracea* var. *botrytis* L.), sprouting broccoli (*Brassica oleracea* var. *italica*), and rutabagas (*Brassica napus* or *napobrassica*). The flavor of these vegetables is blander than that of members of the allium family although sulfur compounds still play an important role. Unlike the vegetables of the allium family, these vegetables are normally cooked before consumption.

A. Sulfides

The *S*-methyl-L-cysteine sulfoxide isolated from cabbage by Synge and Wood (1956) accounted for approximately 25% of the total organic sulfur present. Dateo *et al.* (1957) established the presence of H_2S and dimethyl disulfide among the volatiles of cooked cabbage using a series of precipitation traps. The formation of dimethyl disulfide was thought to originate from *S*-methyl-L-cysteine sulfoxide and was incorrectly considered to be one of the major flavor components. MacLeod and MacLeod (1968) however were unable to detect sufficient amounts of the disulfide in the volatiles of cooked cabbage to consider it an important flavor component. Nevertheless, these workers suggested that the presence of such large amounts of the substituted cysteine sulfoxide warranted further research to elucidate its role in the production of cabbage flavor. (MacLeod, 1970).

The chemical composition of the flavor volatiles appeared to vary little among members of the brassica family. For instance, dimethyl monosulfide, the most abundant volatile in cabbage, accounted for 26% of the total organic volatiles. This compound was present at an equivalent level in cauliflower but represented only 10% of the total organic volatiles in

Brussels sprouts. In general, the close similarity among members of this family made it extremely difficult to identify distinctive flavor characteristics.

B. Isothiocyanates

The formation of isothiocyanates from thioglucosides appears to play a prominent role in the flavor genesis of members of the brassica family. Jensen *et al.* (1953), using paper chromatography, discovered the presence of allyl isothiocyanate in white and red cabbage. This extremely pungent allyl derivative was responsible for the aroma of brown mustard and horseradish and partially responsible for the fresh flavor of cabbage. The chemical structure of its precursor thioglucoside in cabbage, sinigrin, was proposed by Ettlinger and Lundeen (1956) to be the following:

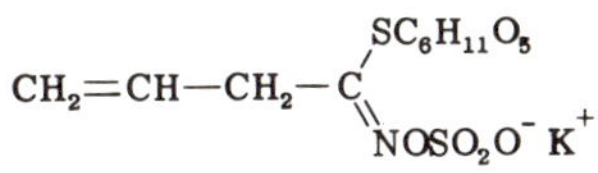

Challenger (1953) suggested that these thioglucosides undergo enzy matic hydrolysis to yield the pungent isothiocyanate. The enzyme involved is myrosinase (EC 3.2.3.1), which catalyzes the following reaction:

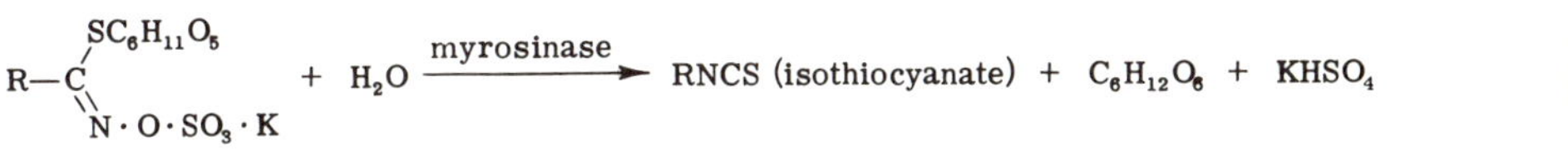

This reaction occurs only after cellular disruption because the enzyme and substrate are compartmentalized separately within the intact cell. The actual reaction occurs in two steps since the product of enzymatic hydrolysis is a thiohydroxamic *O*-sulfate. This compound is extremely unstable and undergoes spontaneous degradation involving the release of sulfate followed by Lossen rearrangement to form the isothiocyanate (Schwimmer and Friedman, 1972):

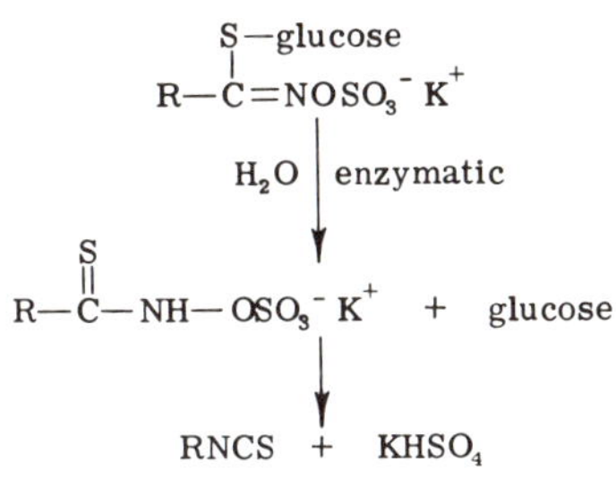

Bailey *et al.* (1961) analyzed the volatile sulfur compounds from fresh cabbage by a combination of gas chromatography and mass spectrometry and identified 20 compounds. These included five isothiocyanates, five sulfides, nine disulfides, and one trisulfide. In contrast to fresh cabbage, dehydrated cabbage samples were almost devoid of these volatiles. Addition of myrosinase to the dehydrated cabbage samples produced allyl, *n*-butyl, and 3-butenyl isothiocyanates, which indicated the stability of the corresponding thioglucoside precursors to the dehydration process (Table 4.3).

The presence of allyl isothiocyanate in fresh cabbage leaves was first reported by Jensen *et al.* (1953) and subsequently confirmed by Bailey *et al.* (1961) and Clapp *et al.* (1959). An additional five isothiocyanates were also reported by Bailey *et al.* (1961), among which *n*-butyl, 3-butenyl, and 3-methylthiopropyl derivatives had been discovered previously in the seeds of other natural products. The predominance of allyl isothiocyanate in fresh shredded cabbage was established by Clapp *et al.* (1959), as indicated in Table 4.4. Bailey *et al.* (1961) demonstrated the central role of allyl isothiocyanate by successfully regenerating this volatile in dehydrated cabbage following treatment with myrosinase. The regeneration of cabbage flavor was also investigated by Schwimmer (1963), MacKay and Hewitt (1959), Hewitt *et al.* (1953), and Konigsbacher *et al.* (1959). All these investigations confirmed the action of myrosinase on sinigrin to produce allyl isothiocyanate, the essential volatile for the flavor of cabbage.

MacLeod and MacLeod (1970a) investigated the flavor volatiles of dehydrated cabbage, since dehydration was the most common technique used for its preservation. The dehydrated product, however, was a poor substitute for the fresh vegetable, since it developed an objectionable

TABLE 4.3

Volatile Isothiocyanates of Fresh and Dehydrated Cabbage[a]

Isothiocyanate	Fresh cabbage	Dehydrated cabbage	Dehydrated cabbage + enzyme
Methyl	+	0	0
n-Butyl	+	0	+
Butenyl	+	0	+
Allyl	+	0	+
Methylthiopropyl	+	0	+

[a] Adapted from Bailey *et al.* (1961).

TABLE 4.4

Estimated Amounts of Isothiocyanates in Fresh Cabbage[a]

Isothiocyanate	Amount (ppm)
Allyl	2.9
3-Methylthiopropyl	0.15
3-Butenyl	0.016
3-Methylsulfinylpropyl	Trace

[a] Reprinted with permission from Clapp *et al.* *J. Am. Chem. Soc.* **81**, 6278. Copyright by the American Chemical Society.

odor when cooked and had an unappetizing appearance. Although there were differences in the levels of saturated aldehydes and alcohols, the most marked differences were evident among the sulfur volatiles. With the exception of one sample, the dehydrated cabbage possessed considerably less dimethyl sulfide than the fresh cabbage. Of the allyl compounds the more important ones included allyl isothiocyanate, allyl cyanide, and allyl alcohol. As shown in Table 4.5, a decrease in isothiocyanate was evident in the majority of the dehydrated samples and was accompanied by an increase in the corresponding nitrile. This relationship between isothiocyanate and nitrile indicated the availability of a precursor for the formation of both of these compounds. Whereas the decrease in isothiocyanate could be attributed to the inactivation of myrosinase, the increase in alcohol and nitrile appeared to be nonenzymatic in origin. Consequently, MacLeod and MacLeod (1970a) suggested that allyl cyanide and allyl alcohol were both produced by decomposition of sinigrin by heat process

TABLE 4.5

Allyl Compounds Produced in Dehydrated versus Cooked, Fresh Cabbage[a]

		Dehydrated			
Compound	Fresh standard (%)	Sample A (%)	Sample B (%)	Sample C (%)	Sample D (%)
Alcohol	0.1	4.5	4.0	1.5	1.0
Nitrile	3.0	48.5	33.5	25.0	23.5
Isothiocyanate	6.5	1.5	0.1	0.1	0.1

[a] From MacLeod and MacLeod (1970b). Copyright © by Institute of Food Technologists.

ing. The presence of small amounts of the isothiocyanate was attributed to some enzymatic hydrolysis taking place between harvesting and processing operations.

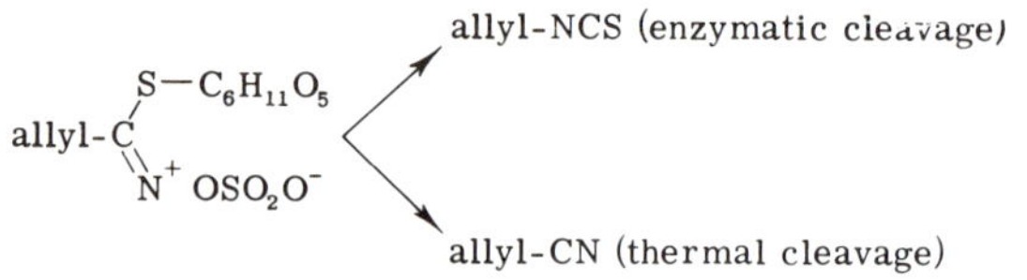

The contribution of different parts of the cabbage to flavor production was also studied by MacLeod and MacLeod (1970b). These workers found that the sulfur volatiles were produced mainly by the young leaves, whereas the older leaves produced greater amounts of aldehydes, ketones, and alcohols. When the whole cabbage was cooked, 36% of the total volatiles were sulfur compounds, of which 46% were contributed by the heart and 15% by the outer leaves. This observation explains why the heart is associated with a stronger flavor than the outer leaves. This was further confirmed by the presence of 36.5% dimethyl sulfide and 16.5% allyl isothiocyanate in the volatiles from the heart compared to 9.5 and 3% respectively, from the outer leaves. The younger leaves also formed other volatiles including ethanol, dipropyl sulfide, allyl cyanide, and methylpropyl disulfide.

The flavor volatiles of several other cooked vegetables were also reported by MacLeod and MacLeod (1970b) including Brussels sprouts and cauliflower. The flavor of sprouts was quite distinct from that of cabbage, which could be related to the smaller amount of sulfur volatiles combined with the absence of unsaturated aldehydes and alcohols, with the exception of *cis*-pent-3-en-1-ol. As was evident with other members of this family, chopping resulted in an increase in isothiocyanates which was accompanied by a stronger and more bitter taste than that of the whole sprouts. The effect of freezing produced several changes, the most prominent of which was a decrease in the formation of allyl isothiocyanate. This was attributed to the destruction of myrosinase during the blanching stage before freezing and was accompanied by an increase in allyl cyanide. Studies on cauliflower were confined to the fresh vegetable, although the total amount of volatiles of the cooked vegetables was far greater than that of cooked cabbage. This was due to the formation of volatiles by the inflorescence rather than the outer leaves. The relative amounts of dimethyl sulfide paralleled those in cabbage, although there was a twofold increase in acetaldehyde compared to the level in cabbage.

Chapter 5

Terpenoids and Flavonoids

I. INTRODUCTION

Terpenes, the major compounds in citrus oils, contribute to the flavor of citrus fruits. (+)-Limonene, a monoterpene hydrocarbon possessing a lemonlike odor, accounts for approximately 90% of most citrus oils. However, it is the oxygenated terpenes or terpenoids, representing only 5% of the oil, that provide the characteristic flavor of individual citrus species. Of these, neral and geranial contribute to the distinctive flavor of lemons, and the bicyclic sesquiterpene nootkatone contributes to the flavor of grapefruit juice (W. D. MacLeod, 1966). These compounds undergo structural changes and hydration in the presence of air or dissolved oxygen. This explains why citrus juice concentrates prepared by low-temperature vacuum evaporation techniques are superior in flavor to those processed at high temperatures.

Flavonoids, which comprise a C_6–C_3–C_6 class of compounds, are also present in citrus fruits. Examples of flavonoids include hesperidin, which is present in sweet oranges, mandarins, and lemons, and naringin, which is found in grapefruit and Seville oranges. Hesperidin is quite tasteless, whereas naringin has an extremely bitter flavor. Its bitterness is acceptable at moderate levels in fruits such as Seville oranges. This fruit makes excellent marmalade because naringin imparts an attractive astringency to the product. This chapter discusses the biosynthesis of the terpenoids and flavonoids that are important to citrus flavor as well as methods for removing components that have an adverse effect on the flavor of citrus products.

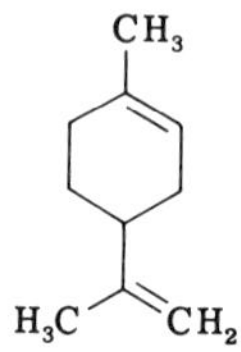

Limonene

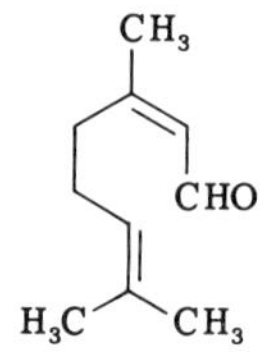

Neral

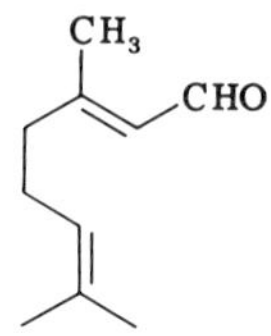

Geranial

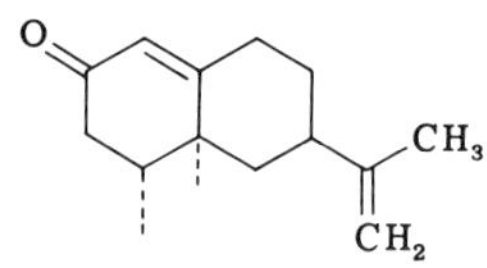

Nootkatone

II. BIOSYNTHESIS OF TERPENOIDS

The biosynthesis of mevalonic acid in oranges was investigated by Potty (1969) because it is a recognized pathway to terpenoid metabolism in plants. Vesicular preparations from mature oranges exhibited both acetyl-CoA synthetase (EC 6.2.1.1) and acetoacetyl-CoA thiolase (EC 2.3.1.9) activity, indicating an ability to synthesize both acetyl-CoA and acetoacetyl-CoA, respectively. The absence of hydroxymethylglutaryl-CoA synthetase (EC 4.1.3.5) or the corresponding reductase (EC 1.1.1.34) indicated that mevalonic acid could not be formed from acetate as reported in other plant systems (Scheme 5.1). Nevertheless, Potty and Bruemmer (1970a) incubated mevalonic acid with orange extracts in the presence of ATP and demonstrated the formation of both mevalonate 5-phosphate (MVA-5P) and mevalonate 5-pyrophosphate (MVA-5PP) by their respective kinases. This suggested that an alternative pathway must be present in the orange by which mevalonic acid is formed. The formation of MVA-5P and MVA-5PP represents the activation step in terpenoid biosynthesis reported previously in liver, yeasts, and plants (Bloch *et al.*, 1959; Beytia *et al.*, 1969; Witting and Porter, 1959). The decarboxylation and dehydration of mevalonate to isopentyl pyrophosphate (IPP) by 5-pyrophosphate-mevalonate carboxylase (EC 4.1.1.33) was established in the orange by Potty and Bruemmer (1970c) using [1-^{14}C]mevalonate. A second enzyme, isopentyl pyrophosphate Δ^3,Δ^2-isomerase, was also identified in orange extracts since 40% of labeled IPP appeared as dimethylallyl pyrophosphate (DMAPP). These reactions are outlined in

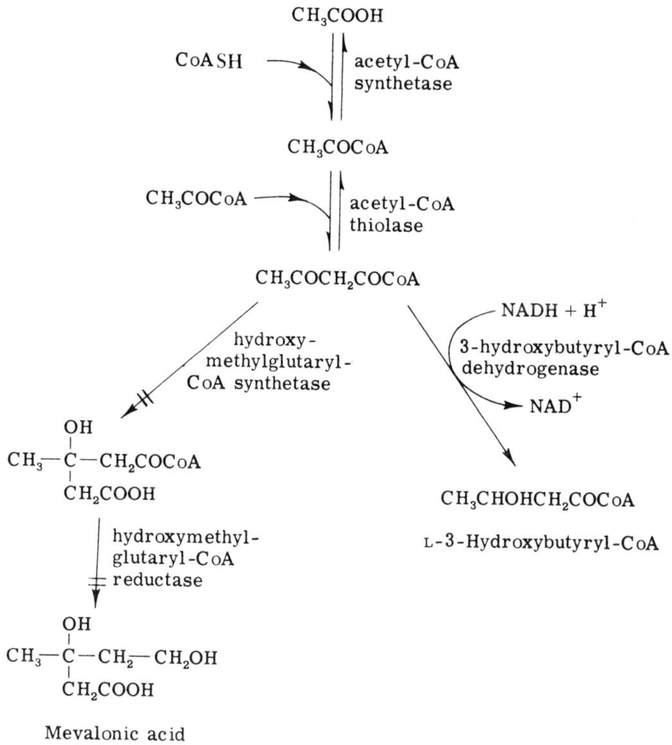

Scheme 5.1. Biosynthesis of mevalonic acid in the orange.

Scheme 5.2. The C_5 compounds IPP and DMAPP condense head to tail to form the various acyclic and cyclic monoterpenes.

A. Acyclic Monoterpenes

Potty and Bruemmer (1970c) reported linalool to be the major monoterpene formed when [2-^{14}C]mevalonate was incubated with orange juice vesicles. These researchers suggested that terpenoid synthesis in citrus was via linaloyl pyrophosphate. The ability of juice vesicle preparations to phosphorylate [3-^{14}C]linalool in the presence of ATP and Mg^{2+} was also reported by Pott and Breummer (1970c). Whether it was a specific kinase or was involved in the activation of mevalonic acid was unclear. However, the identification of a specific kinase for geraniol in peas by Van Aller and Nes (1968) supports the presence of a specific kinase for linalool. The for-

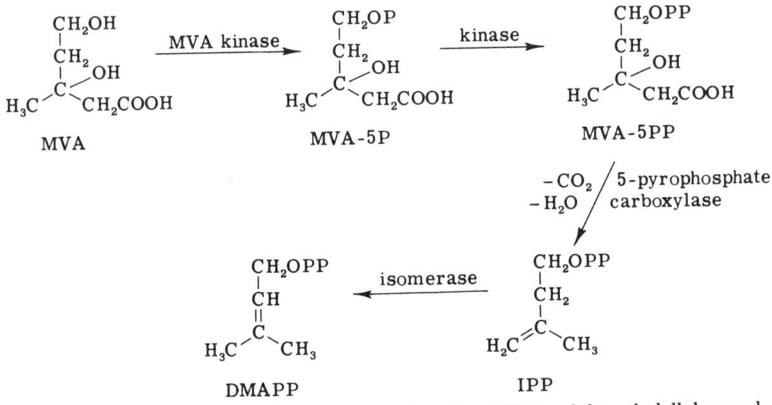

Scheme 5.2. Biosynthesis of isopentyl pyrophosphate (IPP) and dimethylallyl pyrophosphate (DMAPP) from mevalonic acid (MVA) in the mature orange.

mation of linalool appeared to be unique to the citrus fruit, since most plant and animal systems form geraniol pyrophosphate. Attaway *et al.* (1967) proposed the sequence shown in Scheme 5.3 to explain the formation of linaloyl pyrophosphte from IPP and DMAPP involving a hypothetical enzyme, linaloylpyrophosphate synthetase. George-Nascimento and Cori (1971) incubated [2-^{14}C]mevalonic acid with cell-free extracts of

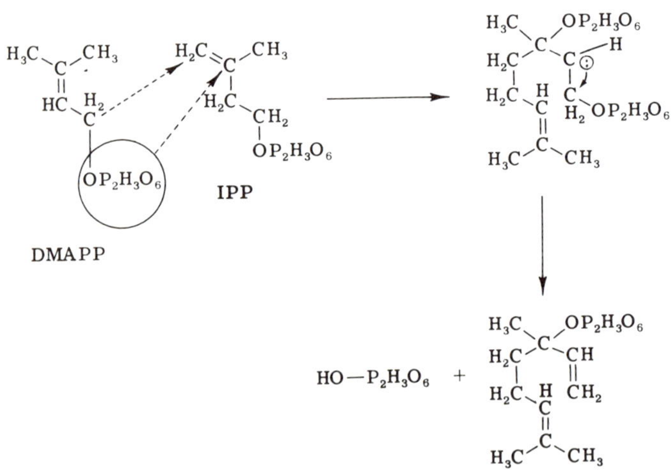

Scheme 5.3. Possible biosynthesis of linaloyl pyrophosphate (Attaway *et al.*, 1967).

orange flavedo and reported the production of both nerol and geraniol pyrophosphates. The discrepancy between these results and those of Potty and Bruemmer (1970c) reflect differences between the flavedo and orange juice vesicles. The flavedo, a rich source of terpenoids, accumulated much larger amounts of geraniol and nerol pyrophosphates than did the juice vesicles. This appears to explain the inability of Potty and Bruemmer (1970c) to detect either geraniol or nerol pyrophosphates in their juice vesicles. The condensation of IPP and DMAPP to neryl and geranyl pyrophosphates is shown in Scheme 5.4.

B. Cyclic Monoterpenes

The possible role of linaloyl pyrophosphate as the acyclic precursor of cyclic monoterpenes in oranges was suggested by Attaway *et al.* (1967). These researchers (Attaway *et al.*, 1967) found a direct correlation between linalool and limonene levels in citrus cultivars. During maturation of oranges and tangarenes, a decrease in the level of linalool was observed in the peel oils accompanied by an increase in (+)-limonene concentration, as shown for tangarenes in Fig. 5.1. Attaway and Buslig (1969) reported the cyclization of [3-^{14}C]linaloyl pyrophosphate to α-terpineol in

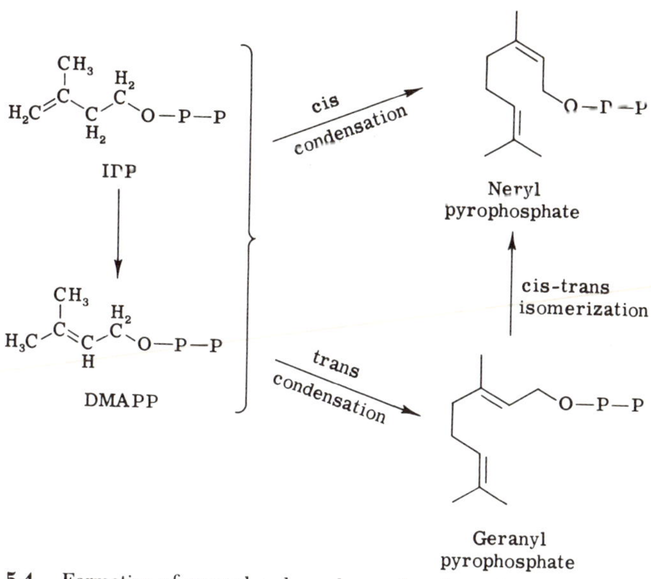

Scheme 5.4. Formation of geranyl and neryl pyrophosphates from IPP and DMAPP.

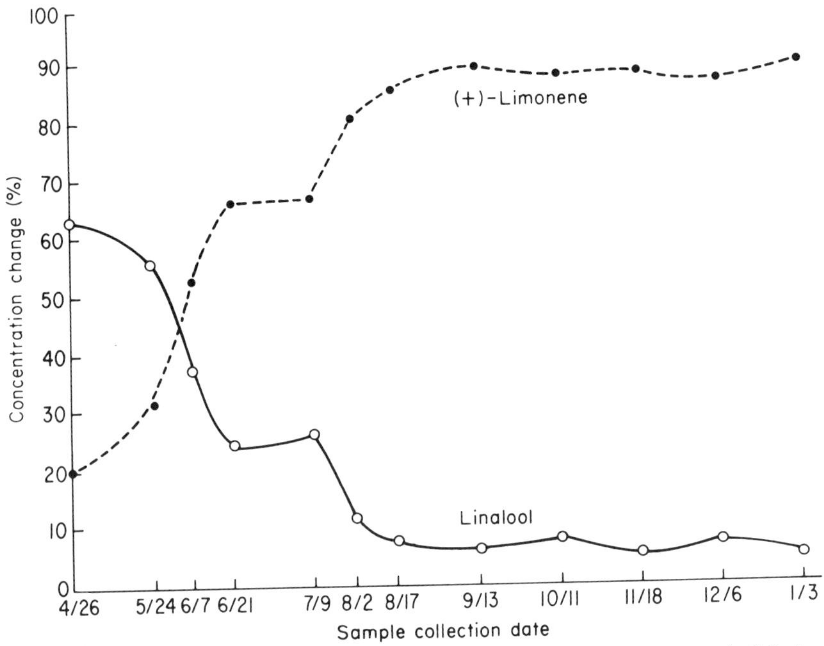

Fig. 5.1. Relative percentages of (+)-limonene and limalool in tangarene peel oil during 1965–66 season (Attaway *et al.*, 1967).

young lemon leaves (*Citrus jambruish*). The product α-terpineol has a characteristic lemon odor (Furia and Bellanca, 1975). Attaway and Buslig (1969) also demonstrated the incorporation of [3-^{14}C]linalool into tertiary cyclic alcohols such as α-terpineol and terpinen-4-ol in rough lemon leaf and tangerine peel oils. Their results (Table 5.1) show that tangerine peel oil had significant activity in α-terpineol as well as some activity in trans2,8 p-menthadiene-1-ol, terpinen-4-ol, limonene and hydrocarbons. *trans-p*-Menthadien-1-ol is a normal cyclic terpenoid found in orange and tangerine fruit. Croteau and Loomis (1975) criticized the long incubation period used in these studies, during which time considerable degradation of the precursor may have occurred. They also noted that α-terpineol had not been degraded in order to determine the position of the label.

George-Nascimento and Cori (1971) incubated isopentyl [4-^{14}C]pyrophosphate with cell-free extracts of orange flavedo and reported the formation of phosphorylated derivatives of dimethylallyl, nerol, geraniol, linalool, and α-terpineol. The ability of orange preparations to form α-

TABLE 5.1

Incorporation of [3-^{14}C]Linalool into Volatile Components of Tangarene[a]

Compound	Radioactivity (cpm)
α-Pinene	0
β-Pinene	45
Sabinene	—
Myrecene	45
α-Terpinene	—
(+)-Limonene	286
cis-Ocimene	—
γ-Terpinene	222
Octanal	85
Terpinolene	85
Methylheptenone	—
Unidentified	—
Unidentified	34
Linalool	8610
Unidentified	—
Terpinen-4-ol	430
trans-2,8-*p*-Menthadien-1-ol	570
β-Caryophyllene	—
α-Terpineol	1350
Citral	88
Geraniol	102

[a] From Attaway and Buslig (1969).

terpineol pyrophosphate or linaloyl pyrophosphate provided further evidence regarding their possible role as intermediates in terpenoid synthesis. These researchers also reported the conversion of neryl and geranyl pyrophosphates to limonene, whereas pinene was formed only in the presence of the *cis* isomer, neryl pyrophosphate. They suggested that limonene might be formed by two distinct pathways or by isomerization of geranyl and neryl pyrophosphates. The absence of neryl pyrophosphate in systmes incubated with geranyl pyrophosphate tended to rule out isomerization. Croteau and Loomis (1975), in reviewing these results, could explain the conversion of geranyl pyrophosphate to limonene only on the basis of isomerization. They suggested that isomerase and cyclase were probably tightly bound so that any isomerization of neryl pyrophosphate from geranyl pyrophosphate would remain undetected due to its immediate cyclization to cyclic monoterpenes. The generally accepted mechanism for the formation of *p*-menthane monoterpenes from neryl and linaloyl pyrophosphates via a carbonium ion is shown in Scheme 5.5. The formation of limonene and terpinolene is due to stabilization of the carbonium ion by loss of a proton or capture of a hydroxyl to form α-terpineol.

C. Interconversion of Acyclic Terpenoids

The typical aroma of citrus fruits is attributed to terpenoids, which represent about 5% of citrus oils (Stanley, 1948). Included in this group are the acyclic isoprenoid alcohols geraniol, nerol, citronellol, and farnesol, which are all present in citrus essence together with their corresponding aldehydes. Orange and grapefruit vesicles were reported to contain substantial amounts of NAD/NADP in both the reduced and oxidized states (Bruemmer, 1969). Potty and Breummer (1970b) examined whether the oxidoreductase systems in citrus fruits were capable of

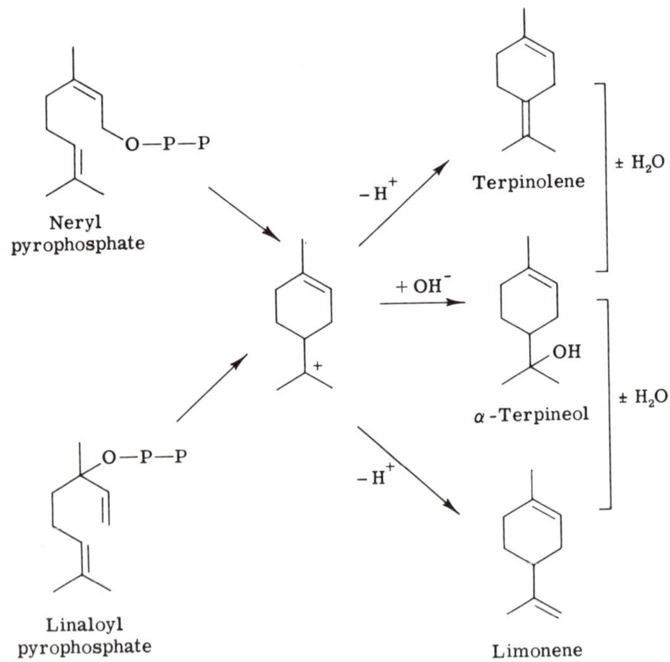

Scheme 5.5. Formation of cyclic monoterpenes from neryl and linaloyl pyrophosphates (Croteau and Loomis, 1975).

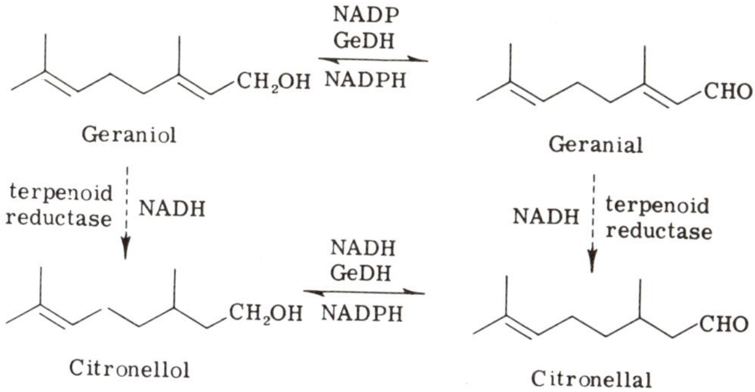

Scheme 5.6. Relationship between acyclic terpenoids in the orange (GeDH, geraniol dehydrogenase) (Potty and Bruemner, 1970b).

mediating the interconversion of these acyclic aldehydes and alcohols. They identified an enzyme system in orange juice vesicles that oxidized geraniol in the presence of NADP. This enzyme, geraniol dehydrogenase, was separated from alcohol dehydrogenase by ammonium sulfate precipitation. It could also oxidize at a slower rate nerol, farnesol, and citronellol. Scheme 5.6 was proposed to explain the interconversion of alcohols and aldehydes in oranges. Breummer (1969) reported that the ratio of NADPH to NADP in orange juice remained relatively constant (1.3:1.5) during maturation. This indicated a high redox potential favoring the reduction of geranial to geraniol. The synthesis of terpenes was probably regulated by the redox potential at the site of formation of these compounds.

Potty and Bruemmer (1970b) found that oxidized glutathione (1×10^{-4} *M*) inhibited geraniol dehydrogenase activity. They attributed this to the glutathione reductase in the enzyme preparation oxidizing any NADPH formed. Limonene was also found to completely inhibit NADPH formation, whereas linalool and α-terpineol exerted a 57% inhibition when present in similar concentrations (1×10^{-4} *M*). All these terpenoids oxidized NADH as well as NADPH in the presence of a suitable enzyme preparation. A limonene reductase system was subsequently identified in the orange by Potty and Bruemmer (1970d) which reduced (+)-limonene to *p*-menth-1-ene in the presence of either NADPH or NADH.

III. LIMONIN AND DELAYED BITTERNESS

A serious problem for the citrus industry is the development of bitterness in the juice of certain varieties of sweet oranges and grapefruit. This phenomenon is referred to as delayed bitterness since it is not observed either in the fresh fruit or in the juice following extraction. Juice held at room temperature develops after several hours (or, if heated, within a few minutes) a definite bitterness. This delayed bitterness is characteristic of juice from early-season to midseason Washington navel, Australian Valencia, and Israeli Shamouti oranges as well as from Marsh grapefruit. The juice obtained from late-harvested fruit from these same varieties, however, does not develop this bitterness.

A. Bitter Component

The bitter component in navel and Valencia oranges was first reported by Higby (1938) and later identified as limonin. In addition to limonin, other related bitter principles that may contribute to citrus juice flavor have been designated limonoids (Kefford, 1959). The structure of limonin was characterized by Arigoni *et al.* (1960) as that of a triterpenoid dilactone with a furan-ring side chain and A- and D-rings both δ-lactones. Limonin, the only limonoid present in bitter navel orange juice (Emerson, 1949), is also found in grapefruit juice (Maier and Dreyer, 1965). Its economic importance stems from its development from a nonbitter precursor, thus preventing the marketing of certain high-quality citrus crops in the form of juices.

Barton *et al.* (1961) demonstrated that the A and D lactone rings in limonin could be reversibly opened with alkali. This suggested that the hydrolyzed disalts or monolactones could act as limonin precursors. Maier and Beverly (1968) studied the capillary membranes, albedo, juice vesicles, and seeds of early season Washington navel oranges and Marsh grapefruit and found that limonin was present only in the juices that developed bitterness. No limonin was detected in the fresh fruit to any significant degree, and the major limonoid present in the albedo and endocarp tissues was limonin monolactone. The monolactone itself was not bitter, but when it came in contact with the acidic citrus juice lactonization occurred, resulting in the conversion of the monolactone to the bitter compound limonin (Scheme 5.7). The stability of the monolactone was attributed to the relatively neutral pH of the tissues in which it was located. The absence of delayed bitterness in the juice from the late-season fruit was explained by the disappearance of limonin in the capillary tissues. Although two monolactone forms are possible, Maier and Margileth (1969) established the A-ring monolactone as the only naturally occurring one.

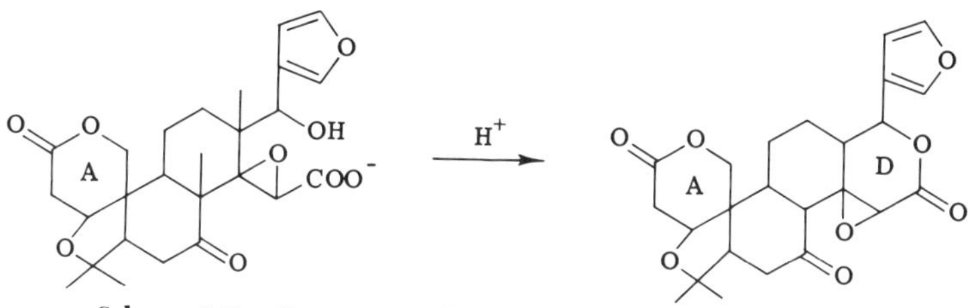

Scheme 5.7. Conversion of limonin A-ring lactone to limonin.

B. Metabolic Debittering

Flavian and Levi (1970) demonstrated the disappearance of limonin in blended Shamouti orange peel, which was not evident with the corresponding boiled extract. They postulated that a catalytically active limonin degrading system was responsible for the decrease in limonin during maturation. Chandler (1971) confirmed the presence of a limonin degrading enzyme in citrus albedo of navel and Valencia oranges. The reaction was conducted over the pH range 5.4–6.2 and the temperature range of 15°–45°C to obtain the maximal degradation of limonin. These researchers used the dinitrophenylhydrazone method to determine limonin, which was far more specific than the hydroxamate method used by Flavian and Levi (1970). The enzymatic degradation of limonin appeared to be unaffected by the presence of butylated hydroxyanisole, thus confirming its independence from limonin autoxidation to evodol and limonexic acid (Barton *et al.*, 1961; Melera *et al.*, 1957).

Maier and Margileth (1969) reported the presence of a metabolic debittering system in late-season citrus fruits which eliminated bitterness by removing the precursor limonoate A-ring lactone. Consequently, little or no limonin was formed in the juice following extraction. Maier *et al.* (1971) suggested that if a method could be devised to activate the natural debittering system at an earlier stage of fruit maturity then the problem of limonin might be eliminated. Several methods were subsequently investigated by these researchers including the application of the so-called ripening hormone ethylene, and CEPA (2-chloroethylphosphoric acid). The latter compound is known to generate ethylene, which in turn accelerates limonoid metabolism. Maier *et al.* (1971) clearly demonstrated accelerated limonoid metabolism in navel oranges in response to ethylene and CEPA treatment as shown by the decreased limonin content in the juice (Table 5.2).

A more detailed study of the effect of ethylene on limonoid metabolism in citrus fruits was carried out by Maier *et al.* (1973). Exposure of navel

TABLE 5.2

Effect of CEPA and Ethylene Treatment of Navel Oranges on Limonin Production[a,b]

Treatment	Limonin in juice (ppm)	Decrease (%)
Zero	12.4	0
Control	10.8	13
CEPA	8.4	32
Continuous ethylene	9.0	27

[a] From Maier *et al.* (1971).
[b] The fruit was held for 8 days.

oranges to 20 ppm ethylene for 3 hours was as effective in accelerating limonoid metabolism as exposure at the same level for 20 hours or continuously for 5 days (Table 5.3). The internal ethylene concentration of the fruit exposed for 3 hours rose rapidly to 17 ppm during the exposure period and then decreased rapidly with removal of the exogenous ethylene source. This indicated that within 48 hours the internal ethylene concentration was identical to that of the untreated fruit. A similar situation was observed for carbon dioxide levels, which after an initial rise paralleled those of the untreated fruit over the remaining 5 days of storage. In contrast, however, the fruits subjected to the 5-day ethylene treatment retained a constant ethylene level of 17 ppm throughout ac-

TABLE 5.3

Effect of Time of Exposure to (20 ppm of) Ethylene on Percent Loss of Limonin in Naval Oranges[a]

Treatment	Days held at 70°F	Decrease in juice limonin (%)
Untreated	0	0
Untreated[b]	5	14
3-Hour ethylene[b]	5	32
20-Hour ethylene[b]	5	31
5-Day ethylene[b]	5	36

[a] Reprinted with permission from Maier *et al. J. Agric. Food Chem.* **21**, 490, 1973. Copyright by the American Chemical Society.
[b] Total time from initiation of treatment.

companied by a fourfold increase in CO_2 levels. These results suggested that a short ethylene treatment was sufficient to activate the metabolic debittering system without an acceleration in respiration. The relationship between limonin content and bitterness in navel orange juice was evaluated by a taste panel after sufficient time was allowed for the development of full bitterness. The most preferred juice was produced by the 3-hour ethylene treatment and a holding temperature of 70°F even though the 20-hour ethylene treatment and holding temperature of 86°F were associated with the lowest limonin content (Table 5.4). The low preference and high bitterness rating reported for the latter juice was due to the presence of off-flavors which obscured taste judgments of bitterness. These results confirmed the advantage of short-term treatment for fruits with ethylene (3 hours), which directly affected limonoid metabolism compared to the dramatic metabolic and physiological effects associated with the prolonged treatment.

Ethylene treatment also proved to be extremely effective in accelerating limonoid metabolism in grapefruit. A 3-hour exposure to 20 ppm ethylene was as effective as continuous ethylene treatment in reducing the limonoid content, and the levels of naringin and ascorbic acid were unaffected.

TABLE 5.4

Relation of Limonin to Bitterness in Navel Orange Juice[a]

Treatment	Juice limonin (ppm)	Taste evaluation
Untreated, zero time	25.2	Most bitter, least preferred
Untreated, (50°F)[b]	23.6	↓
3-Hour ethylene (50°F)[b]	19.1	
20-Hour ethylene (50°F)[b]	18.2	
Untreated (70°F)[b]	19.8	
20-Hour ethylene (70°F)[b]	14.0	
20-Hour ethylene (86°F)[b]	8.4	
Untreated (86°F)[b]	13.6	
3-Hour ethylene (86°F)[b]	10.8	
3-Hour ethylene (70°F)[b]	13.0	Least bitter, most preferred

[a] Reprinted with permission from Maier *et al.* *J. Agric. Food Chem.* **21**, 490, 1973. Copyright by the American Chemical Society.

[b] Fruit was held 20 hours at 70°F including treatment time, then held 5 days at the indicated temperature.

These workers proposed the short-term ethylene treatment as a simple, effective, and inexpensive procedure for reducing the levels of limonin in extracted juices.

C. Enzymatic Debittering

An alternative method for reducing limonin in extracted juice involved the use of a specific enzyme to degrade limonin. Maier and Beverly (1968) and Maier and Margileth (1969) attributed delayed bitterness to the conversion of lactone to limonin in the citrus juice. This suggested that the metabolism of the lactone could control or inhibit limonin development in the fruit. Hsu *et al.* (1973) isolated 17-dehydrolimonoate A-ring lactone from orange peel and juice as well as from lemon seeds and seedlings. This compound appeared to be the initial product of limonoate A-ring lactone metabolism in citrus fruits. An enzyme, limonoate dehydrogenase, was isolated from *Arthrobacter globiformis* (Hasegawa *et al.*, 1972) and *Pseudomonas* species (Hasegawa *et al.*, 1974b) which catalyzed the dehydrogenation of the free 17-hydroxyl group of limonoate A-ring lactone to produce 17-dehydrolimonoate A-ring lactone (Scheme 5.8). The formation of limonoate A-ring lactone itself was attributed by Maier *et al.* (1969) to the presence of limonin D-ring lactone hydrolase in citrus seeds. This enzyme catalyzes the conversion of limonin to limonin A-ring lactone, thus providing the substrate for dehydrogenase attack. Limonoate dehydrogenase activity was subsequently found by Hasegawa *et al.* (1974a) in the albedo tissues of navel oranges. This confirmed the ability of citrus fruits to form 17-dehydrolimonoate A-ring lactone. The interruption of limonoid metabolism in postharvest fruit in the absence of oxygen supported the limonoate dehydrogenase pathway as the one operating in citrus fruit.

Brewster *et al.* (1976) compared limonoate dehydrogenase from *Pseudomonas* and *Arthrobacter* species. The enzyme from *Pseudomonas* sp. 311-18 appeared to be much more stable under the low pH conditions prevalent in navel orange juice and required NAD^+ or NADP as well as sulfhydryl groups for activity. When this enzyme was added at a level of 200 munits/ml of freshly prepared navel orange juice in the presence of 5 μmoles/ml NAD^+ the limonin content decreased from 21.3 to 3.3 ppm in 1 hour. This level is well below the bitterness threshold range of limonin in orange juice established by Guadagni *et al.* (1973). The increase in limonoate dehydrogenase activity associated with NAD^+/NADP necessitated a method for reoxidizing the reduced forms of these cofactors. This

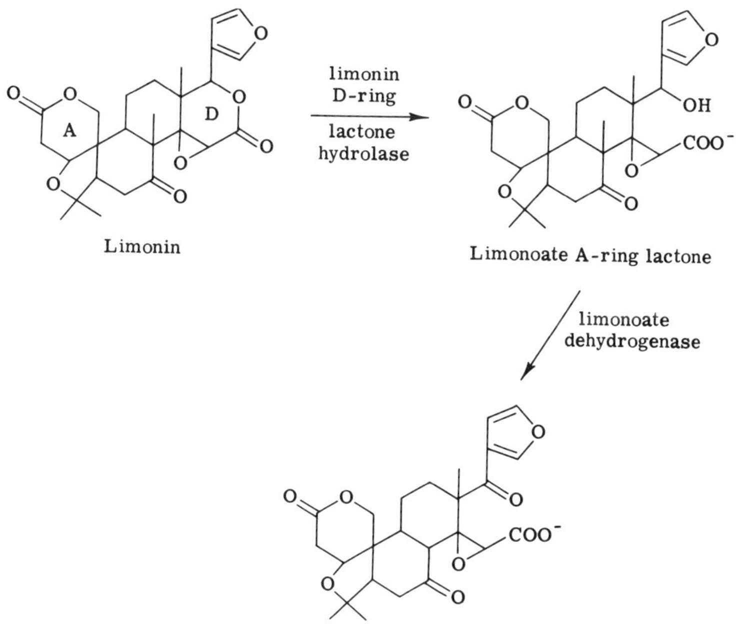

17-Dehydrolimonoate
A-ring lactone

Scheme 5.8. Pathway for enzymatic degradation of limonin (Hasegawa *et al.*, 1972).

was provided by the addition of diaphorase, an enzyme in the electron transport system that is capable of oxidizing $NADH_2/NADPH_2$. Brewster *et al.* (1976) found that 1 unit of diaphorase (Worthington) per milliliter of juice was equivalent to 1 μmole of NAD^+/NADP per milliliter of juice. The efficiency of limonoate dehydrogenase from *Pseudomonas* in decreasing the limonin content in navel orange juice compared to the enzyme from *Arthrobacter* is shown in Fig. 5.2. A substantial decrease in limonin content was evident for the *Pseudomonas* enzyme at all levels of activity, resulting in a 93% reduction of limonin compared to 23% for the corresponding *Arthrobacter* enzyme at an equivalent activity level. This study demonstrated the potential of limonoate dehydrogenase from *Pseudomonas* in minimizing the development of bitterness in citrus juices.

In addition to limonoate dehydrogenase, Hasegawa *et al.* (1974c) isolated another limonoid metabolic system from *Pseudomonas* sp. 321-18. The addition of limonin to this system produced exocellular traces of deoxylimonin together with considerable quantities of deoxylimonoate as

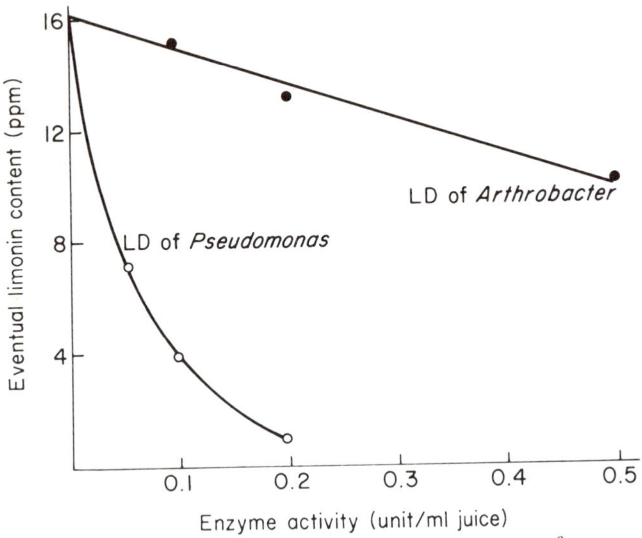

Fig. 5.2. Effectiveness of limonate dehydrogenase (LD) enzymes from two sources in orange juice. The juice was treated at pH 3.5 with LD and 5 μmoles of NAD^+ per milliliter of juice for 2 hours at 24° C. Reprinted with permission from Brewster *et al. J. Agric. Food Chem.* **24**, 21, 1976. Copyright by the American Chemical Society.

metabolites. It thus appeared that limonoate was metabolized via deoxylimonin even though considerable dehydrogenase activity was present. Brewster *et al.* (1976) suggested that limonoate was transformed by hydrolase to limonin, which was then further metabolized to deoxylimonin. Although the enzyme involved in the formation of deoxylimonin remains to be identified, deoxylimonin hydrolase was isolated and characterized by Hasegawa *et al.* (1974c). This enzyme was responsible for converting deoxylimonin to deoxylimonic acid (Scheme 5.9).

In contrast to *Pseudomonas*, when *Arthrobacter globiformis* was grown on a limonin medium, the major metabolite formed was 17-dehydrolimonoate A-ring lactone. It was apparent that *Pseudomonas* sp. 321–18 was capable of metabolizing both limonin and limonoate by two pathways, one through deoxylimonin and the other via 17-dehydrolimonoate or 17-dehydrolimonoate A-ring lactone. Hasegawa (1976) isolated limonin D-ring lactone hydrolase activity from *Pseudomonas* which was similar to the enzyme isolated from grapefruit seeds by Maier *et al.* (1969). Although the pH optima were similar with respect to the hydrolysis of limonin to limonoate A-ring lactone, the citrus enzyme was far more heat stable. Hasegawa (1976) showed that the hydrolase was involved in limonoid degrada-

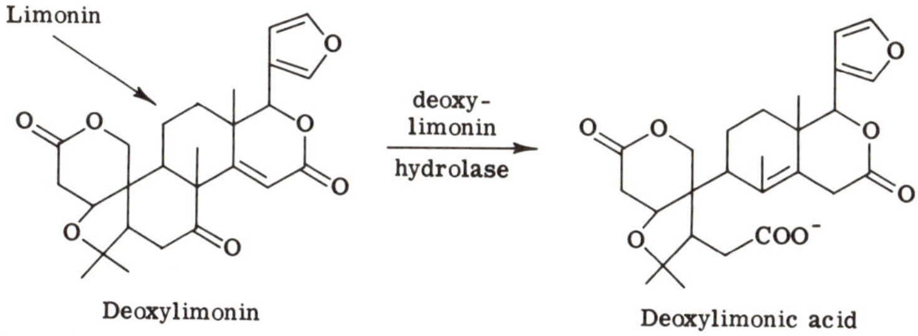

Scheme 5.9. Enzymatic formation of deoxylimonic acid from deoxylimonin. Reprinted with permission from Brewster *et al. J. Agric. Food Chem.* **24**, 21, 1976. Copyright by the American Chemical Society.

tion, although whether this enzyme has a role in limonoid biosynthesis remains to be elucidated. The overall scheme for the metabolism of limonin by these enzymes is shown in Scheme 5.10.

Vaks and Lifshitz (1976) identified an enzyme from *Acinetobacter* sp., a gram-negative rod soil bacterium capable of attacking limonin close to the natural pH of the juice. This enzyme was distinct from those discussed previously and was inactive at pH 8.5, where limonin is in the open-ring limonoate form. However, the mechanism of the reaction catalyzed and the nature of the products formed were not discussed.

D. Limonin Reduction by Chemical Sorbents

The problem of bitterness appears to have been solved through the recent use of chemical sorbents. An early attempt to reduce limonin was reported by McColloch (1950) using activated carbon; however, this material was too nonselective to be of any practical value. More recent interest in chemical sorbents led to an examination of polyamide powders

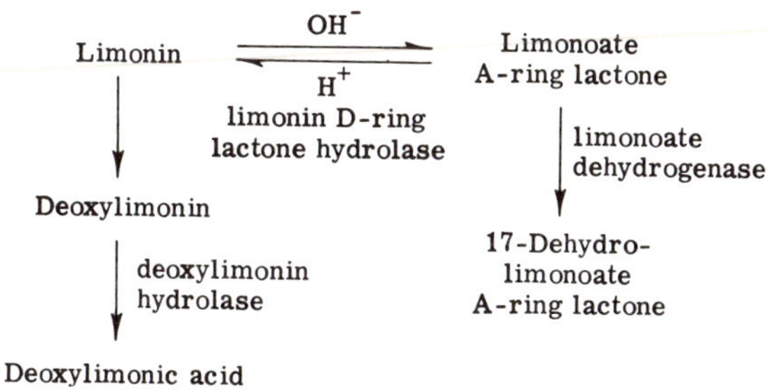

Scheme 5.10. Metabolism of limonin.

by Chandler *et al.* (1968). These researchers found that limonin could be reduced by these powders, although they proved to be more effective in removing flavonoids as well as some ascorbic acid. This lack of specificity was a major drawback in the industrial utilization of these powders. A series of sorbents were examined by Chandler and Johnson (1977), who showed that cellulose acetate was far more specific than polyamide powders, effectively reducing the limonin content of navel orange juice while exerting a negligible effect on flavonoids or ascorbic acid. Figure 5.3 indicates that, although the different samples of orange juice sera varied in their initial levels of limonin, cellulose acetate effectively removed around 80% of the limonin present within 40 minutes. This meant a reduction of limonin in sera from 12–17.2 to 2.0 ± 0.3 mg/liter, which is below the threshold limit reported by Guadagni *et al.* (1973). The only other sorbent material found to be as effective as cellulose acetate was cellulose acetate butyrate. Chandler (1977), in reviewing this bitterness problem, discussed the production of cellulose acetate gel beads that were successfully used in a "gel debittering process." This process provided a simple and inexpensive method for the removal of limonin, thus permitting the commercial use of citrus juices previously affected by the delayed bitterness phenomenon.

E. Biosynthesis of Limonin

The biosynthesis of limonin was demonstrated in germinated seeds of Valencia oranges by Datta and Nicholas (1968). These workers reported

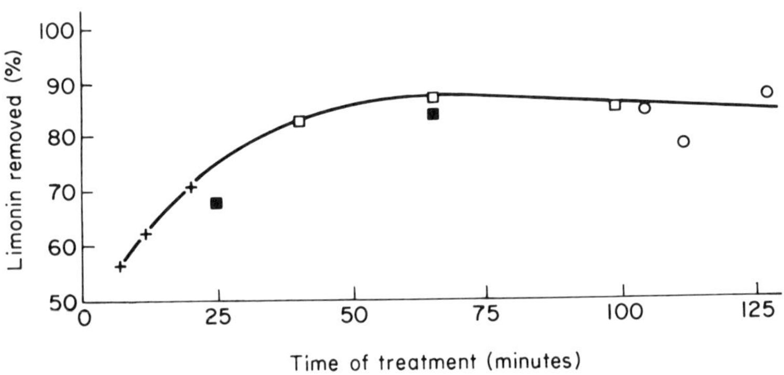

Fig. 5.3. Effect of treatment time on removal of limonin in orange juice sera by cellulose acetate powder (0.210- to 0.355-mm particle size at 26 gm/liter). Initial limonin content of sera milligrams per liter): +, 17.2; ■, 14.0; □, 13.0; ○, 12.0 (Chandler and Johnson, 1977).

the incorporation of [2-^{14}C]mevalonic acid into the triterpene [^{14}C]-limonin, although the low activity precluded it from being a major site of limonoid formation in citrus fruits. The synthesis of limonoids during the growth of citrus fruits was thought by Kefford and Chandler (1970) to occur in the albedo. Hasegawa and Hoagland (1977), however, conducted radioisotope tracer studies using sodium [2-^{14}C]acetate and [2-^{14}C]-mevalonate and reported that limonate A-ring lactone was synthesized in the citrus leaves. The results (Fig. 5.4) indicate an increase in limonoate A-ring lactone during the growth of lemon leaves (when expressed as micrograms per leaf), although the total amount synthesized decreased when expressed as parts per million. These results indicate a greater synthesis of limonoate A-ring lactone in the younger leaves which decreased as the leaf increased in size. This compound was apparently translocated to the fruit, although the possibility of a precursor being formed in the leaves translocated to the fruit and converted to limonoate A-ring lactone could not be ruled out.

Arigoni *et al.* (1960) suggested that the precursor of limonin was a tetracyclic triterpene similar to euphenol, although there was no experimental evidence to support this. Dreyer (1968) proposed the biosynthetic pathway for limonin shown in Scheme 5.11. Obacunone is the second

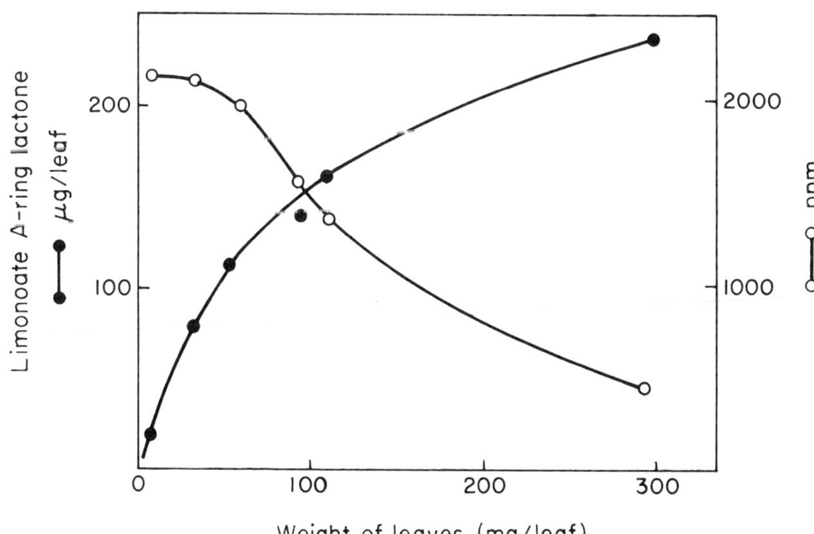

Fig. 5.4. Changes in limonoate A-ring lactone content during the growth of lemon leaves (Hasagawa and Hoagland, 1977).

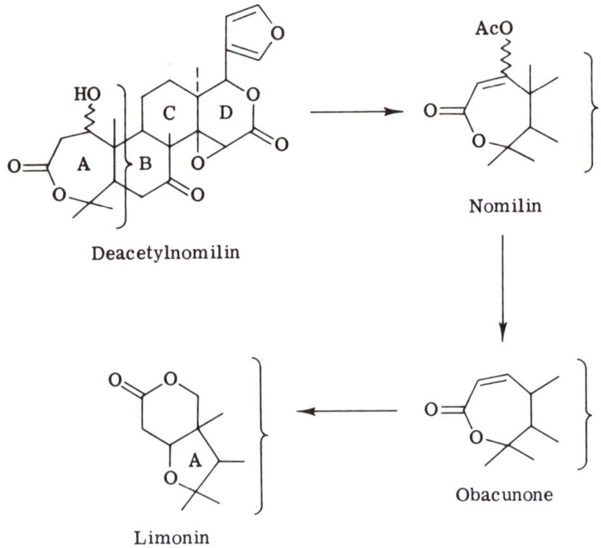

Scheme 5.11. Biosynthesis of limonin.

most prominent citrus limonoid present in all fruits with the exception of sour oranges and limes, and its conversion to limonin was confirmed by Tokoroyama and Matsuura (1962). At least two steps involving obacunoic acid and isoobacunoic acid were considered essential for the conversion of obacunone to limonin (Dreyer, 1968). Bennett (1971) isolated five acidic limonoids from a grapefruit seed extract including isoobacunoic, epiiso-obacunoic, nomilinic, and deacetylnomilinic acids. The last two compounds had not been reported previously and were hydroxy acid derivatives of nomilinic and deacetylnomilinic acids. The presence of isoobacunoic acid in grapefruit seeds supports the hypothesis that it is a precursor of limonin. This acid could be formed from obacunone via obacunoic acid or by cyclization of deacetylnomilinic acid. Since no obacunoic acid was detected in this study the latter pathway appeared to be involved. The conversion of deacetylnomilinic acid to ichangin was reported by Dreyer (1966) as an alternative route to limonin formation in ichangin lemon seeds. The fruit is recognized for its large accumulation of ichangin, that can be converted to limonin. Although our knowledge of the pathway leading to limonin biosynthesis remains incomplete, it nevertheless appears that deacetylnomilinic acid may be the key intermediate. The structures of the acidic limonoids are shown in Scheme 5.12.

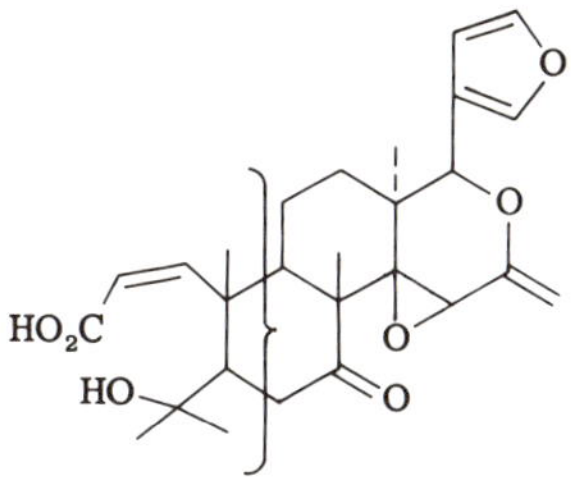

Obacunoic acid

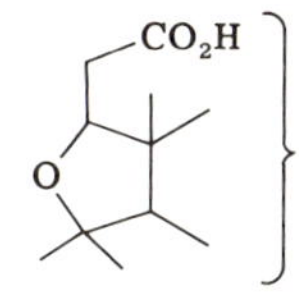

Isoobacunoic acid

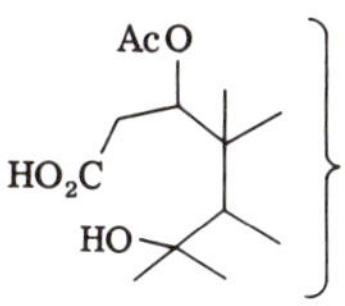

Nomilinic acid

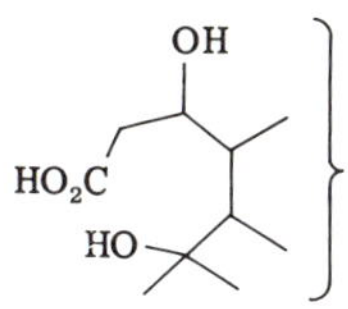

Deacetyl-nomilinic acid

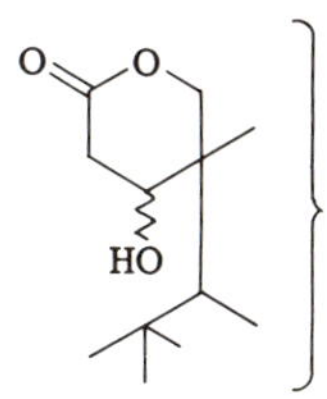

Ichangin

Scheme 5.12. Structures of acidic limonoids.

IV. FLAVONOIDS AND BITTERNESS IN CITRUS FRUITS

The peels of oranges, lemons, and grapefruits contain a wide range of flavonone glycosides. The most common of these are hesperidin and naringin, the former being the main flavonone constituent of oranges and lemons and the latter being the predominant flavonone in grapefruits. Both flavonones are present as the 7-glycosides, and whereas naringin is intensely bitter hesperidin is tasteless. The structure of hesperidin (Scheme 5.13) consists of three parts: L-rhamnose linked α-1,6 to D-glucose, which in turn is linked to the (−)-hydroxy group of the flavonone 2-(S)-hesperitin (Horowitz, 1964). The disaccharide portion, 6-*O*-α-rhamnopyranosyl-D-glucopyranose, is referred to by its trivial name rutinoside. The structure of naringin is thought to be similar to that of hesperidin, differing only in the aglycone moiety naringenin, in which the B-ring substitution pattern is different (Scheme 5.14).

Horowitz and Gentili (1969) determined the configuration of rhamnose in naringin as well as in two other flavonoids, poncirin and neohesperidin, in order to explain the sensory differences between naringin and hesperidin. They found that the linkage between rhamnose and glucose was 1 →

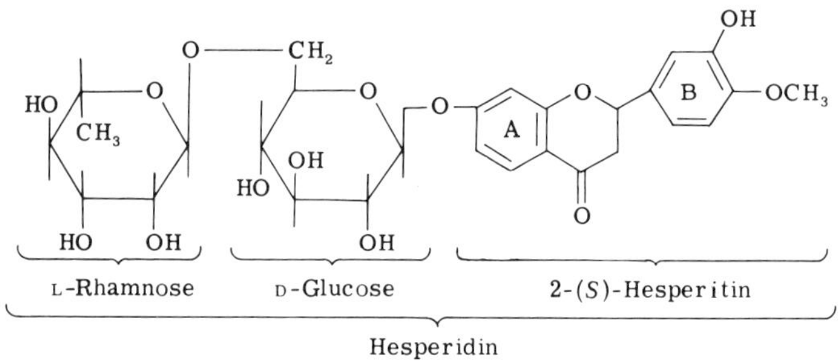

Scheme 5.13. Structure of Hesperidin.

2 in all three flavonoids, suggesting that they could be considered neohesperidases (Scheme 5.15). However, the linkage between rhamnose and glucose in hesperidin was 1 → 6. It was therefore apparent that it was the point of attachment between the rhamnose and glucose that determined the bitterness of the flavonone-7-β-neohesperidosides and the tastelessness of flavonone-7-β-rutinosides.

The presence of naringin at moderate levels is quite acceptable in grapefruit and grapefruit products; excessive amounts, however, have an adverse effect on the products. Consequently, considerable attention has been focused on the biogenesis of naringin in an attempt to control its formation and thereby reduce bitterness in grapefruits.

A. Biogenesis of Narginin

Albach and co-workers (1969) investigated the production of naringin in grapefruit and found that naringenin rhamnoglucosides accumulated throughout the entire growing period from ovary to maturity. The rate of naringin formation, however, appeared to be several thousand times greater during the initial period of fruit development than in the subsequent periods. These workers also observed periodic increases in the production of naringenin glucosides which accompanied growth flushes, suggesting that it might be possible to anticipate increases in bitterness so

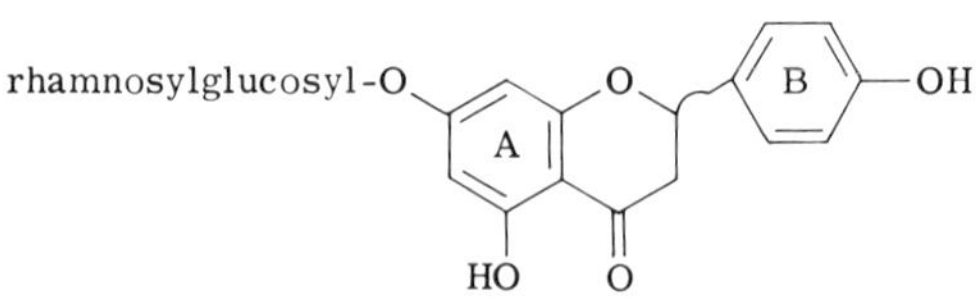

Scheme 5.14. Structure of naringin.

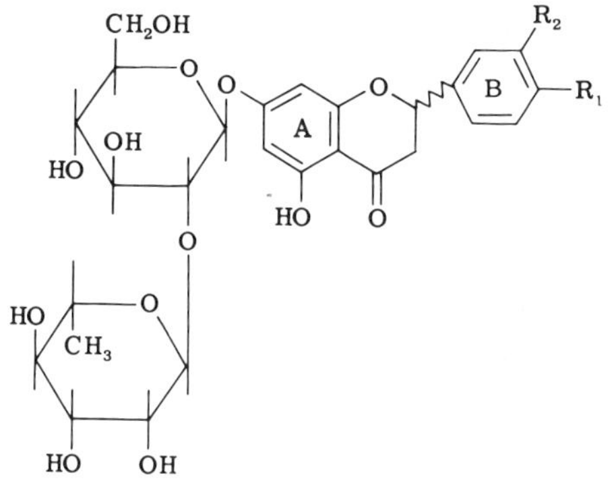

Scheme 5.15. Structures of neohesperidases. Naringin, R_1 = OH; R_2 = H. Poncirin, R_1 = OCH_3; R_2 = H. Neohesperidin, R_1 = OCH_3; R_2 = OH.

that the processing industry could adjust its conditions accordingly. This would ensure the production of grapefruit products of umiform quality. The accumulation of naringin during the initial stages of fruit development appeared to coincide with the formation of new cells. Maier (1969) also reported a rapid increase in naringin during the initial development of Marsh seedless grapefruit. The naringin accounted for 9% of the weight of the fruit (41% of the dry weight) at the end of the first month, leveling off when the fruit had reached one-quarter of its maturity. The peel thickness also reached a maximum when the amount of naringin started to level off, as shown in Table 5.5.

TABLE 5.5

Accumulation of Naringin in Developing Grapefruit[a]

Harvest date	Naringin[b] (gm/fruit)	Peel thickness (cm)	Fresh weight/fruit (gm)
May 5	0.29	0.50	3.21
June 9	1.12	1.17	42.4
July 14	1.45	1.10	121
August 9	1.27	0.80	186
October 13	1.34	0.75	318
December 13	1.53	0.75	443

[a] From Maier (1969).

[b] The method of Fisher *et al.* (1966) was used, and naringin plus naringenin-7-β-rutinoside was determined as a single value. However, because naringin was found to represent over 80% of the total, for simplicity these values are expressed as naringin.

All these studies pointed to the presence of a regulatory system in the cells that was capable of turning on naringin biosynthesis during initial growth and subsequently turning it off. This was thought to be carried out by an inducible enzyme system, as discussed for phenolic compounds by Hanson *et al.* (1967). Fisher (1968) demonstrated the formation of labeled naringin in grapefruit leaves fed L-[^{14}C]phenylalanine. The ability of this amino acid to act as precursor led Maier and Hasegawa (1970) to determine the presence of L-phenylalanine ammonia lyase (EC 4.3.1.5) (PAL) activity in grapefruit tissues during periods of active naringenin glycoside synthesis. This enzyme catalyzes the deamination of phenylalanine to *trans*-cinnamic acid and is the initial step for the biosynthesis of phenolics in plants. The activity of PAL appeared to be highest in the very young fruit and decreased sharply as the size of the fruit increased during the later growth period (Fig. 5.5). The concentration of naringenin glycosides similarly increased during the early growth period of the fruit and slowed

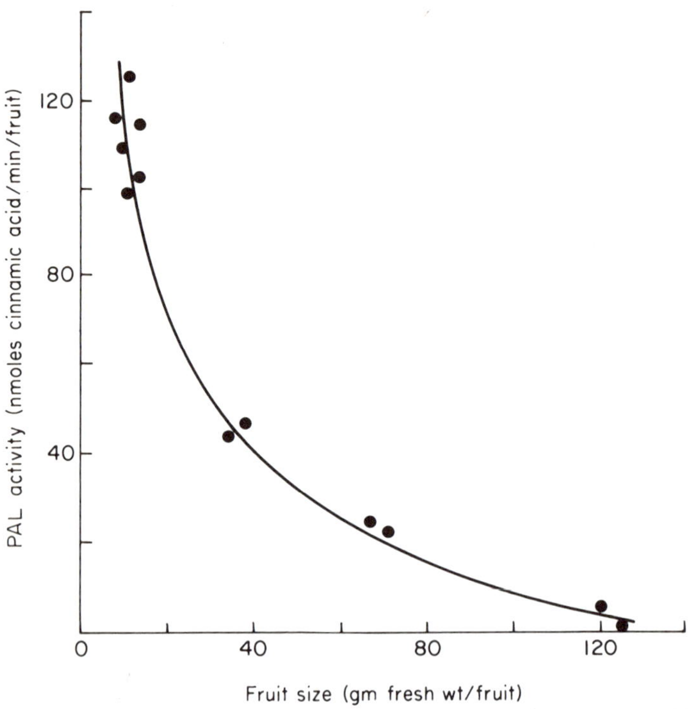

Fig. 5.5. Activity of PAL in developing grapefruit (Maier and Hasegawa, 1970).

down as the fruit matured. A direct relationship between PAL activity and the rate of naringenin production was confirmed by Maier and Hasegawa (1970), suggesting that this enzyme is part of an enzyme–regulator system that controls the synthesis of naringenin glycosides. This confirms earlier observations by Albach *et al.* (1969) and Maier (1969), who found that the increase in naringenin glycosides coincided with extensive cell division during early fruit development rather than with cell enlargement.

Several other enzymes were also identified in young grapefruit (*Citrus paradisi* Macf.) by Hasegawa and Maier (1970). These included L-phenylalanine-2-ketoglutarate amino transferase (EC 2.6.1.5) (PKA) and glutamate dehydrogenase [L-glutamate : NAD(P) oxidoreductase] (EC 1.4.1.2). Both enzymes are present in sufficient amounts to account for the formation of *trans*-cinnamate from phenylalanine and glutamate in *in vitro* studies using cell-free extracts of young grapefruit tissues. The reactions catalyzed are outlined in Scheme 5.16. The formation of phenylalanine from phenyl pyruvate is mediated by PKA in the presence of L-glutamate. The latter acts as an amino donor and is converted to 2-ketoglutarate. Phenylalanine is then catalyzed by PAL to *trans*-cinnamate, and this is accompanied by the release of ammonium ions. L-Glutamate is regenerated in the presence of glutamate dehydrogenase, and the concomitant oxidation of NADH to NAD^+ is achieved by the addition of exogenous alcohol dehydrogenase.

Phenyl pyruvate is recognized as the first aromatic compound produced in the shikimic acid pathway. The absence of tyrosine ammonia lyase activity in grapefruit tissues, reported by Maier and Hasegawa (1970),

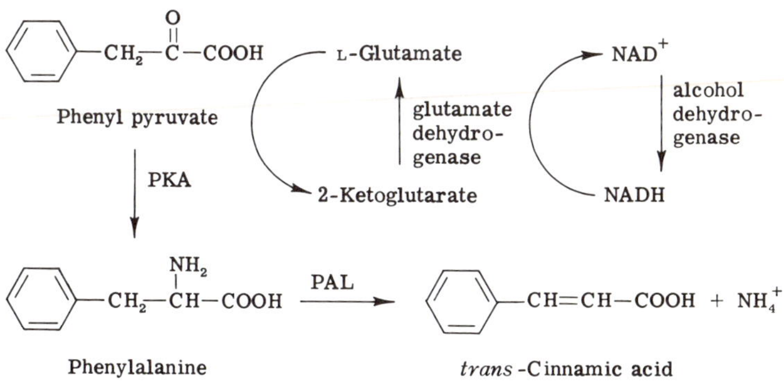

Scheme 5.16. Biosynthesis of *trans*-cinnamic acid (Hasegawa and Maier, 1970).

strongly supported the pathway described as the main one leading to cinnamate production. Using a tissue-slice technique, Hasegawa and Maier (1972) demonstrated for the first time the conversion of ^{14}C-labeled cinnamate to coumarate in grapefruit. The reaction was detected following a 2-hour incubation period; it reached a peak at around 6 hours and then leveled off during the remaining incubation period. A linear increase was observed with respect to the formation of *p*-coumarate with time when the product was expressed on the basis of total radioactivity determined. These results confirmed the presence of an active cinnamate hydroxylase system in the young grapefruit, whereas only traces of activity were observed in the more mature fruit. The decrease in cinnamate hydroxylase activity also corresponded with the decrease in the rate of naringenin glycoside biosynthesis. These researchers observed a high degree of correlation between glycoside accumulation and PAL activity which was not evident with either PKA or glutamate dehydrogenase. The activities of the latter two enzymes remained high even when there was a sharp decline in the rate of glycoside accumulation (Fig. 5.6).

These results indicate the key role of PAL, which may be both regulatory and synthetic, in the biosynthesis of flavonoids in the grapefruit. The absence of phenylalanine hydroxylase and tyrosine ammonia lyase rules out tyrosine as an effective precursor of flavonoid biosynthesis in grapefruit.

The A ring of naringin is formed from the head-to-tail condensation between two malonyl-CoA units and acetyl-CoA, whereas the B ring and the C_3 unit are synthesized from a C_6-C_3 precursor such as cinnamic acid. Conflicting evidence was produced, however, regarding whether hydroxylation of the B ring takes place before or after ring closure of the C_{15} nucleus (Patschke *et al.*, 1966; Sutter and Grisebach, 1969; Amrheim and Zenk, 1969). The isolation by Hasegawa and Maier (1972) of cinnamate hydroxylase in grapefruit discussed earlier indicates that hydroxylation of the B ring occurs before the formation of the C_{15} intermediate. It still remains unclear, however, whether the C_{15} intermediate is a chalcone, a flavonone, or a closely related compound. Current research supports a chalcone, which is probably converted to the flavonone by a specific enzyme, chalcone-flavonone isomerase, which was identified in grapefruit by Maier and Hasegawa (1970). This enzyme was previously reported in the peel of several other citrus fruits (Shimokoriyama, 1957) and more recently in soybean seedlings (*Soya hispida*) (Wong and Moustafa, 1966). The evidence accrued so far indicates the existence of the pathway shown in Scheme 5.17 for the production of naringenin from cinnamic acid.

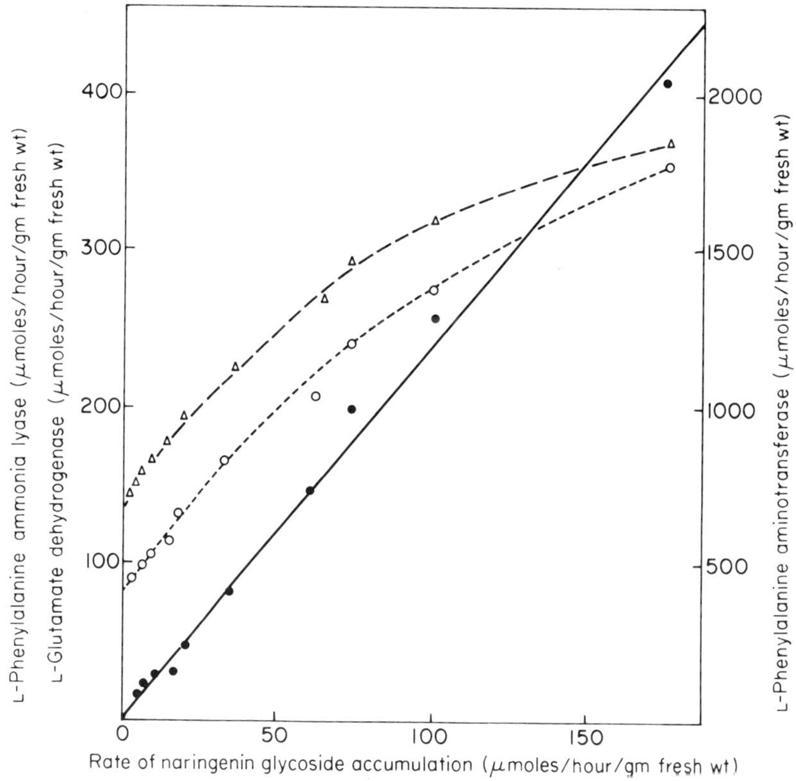

Fig. 5.6. Correlation between activities of PAL, glutamate dehydrogenase, and PKA and the declining rate of naringenin glycoside accumulation in grapefruit (Hasegawa *et al.*, 1972). —●—, PAL; ○, L-Glutamate dehydrogenase, --△--, L-Phenylalanine aminotransferase.

B. Naringin Degradation

A decrease in naringin content was observed by Hagen *et al.* (1966) during the ripening of grapefruit, and this was accompanied by a loss of bitterness. Several mechanisms were proposed including conversion of naringin to the flavonone rhoifolin (Dunlap and Wender, 1962) or to the corresponding 6-rhamnoglucoside (Horowitz, 1964). Both of these compounds are quite tasteless. Maier and Metzler (1967a) isolated dihydrokaempferol from the hydrolyzates of grapefruit flavonoids and suggested that it is an intermediate in the breakdown of naringin to kaempferol (Scheme 5.18). These researchers also identified the flavonone eriodictyol

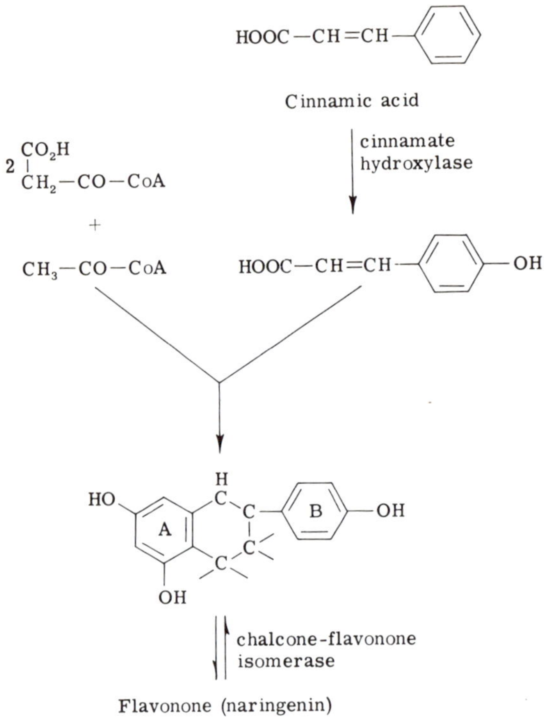

Scheme 5.17. Formation of naringenin from cinnamic acid.

and the corresponding flavonol quercetin together with a number of closely related polyphenols (Maier and Metzler, 1967b). Each compound had hydroxylation and methoxylation patterns similar to those of *p*-coumarate, suggesting *p*-coumarate plays a central role in naringin degradation. Further research is required, however, to delineate these pathways completely.

C. Enzymatic Debittering

The bitter flavor of naringin is associated with the 1 → 2 linkage between rhamnose and glucose discussed earlier. The susceptibility of this glycosidic bond to hydrolytic enzymes suggested the possibility of hydrolyzing naringin to rhamnose and prunin (7-glycoside of flavonone). Prunin can be further hydrolyzed to glucose and naringenin, neither of which is bitter. Ting (1958) added commercial pectic enzyme preparations to

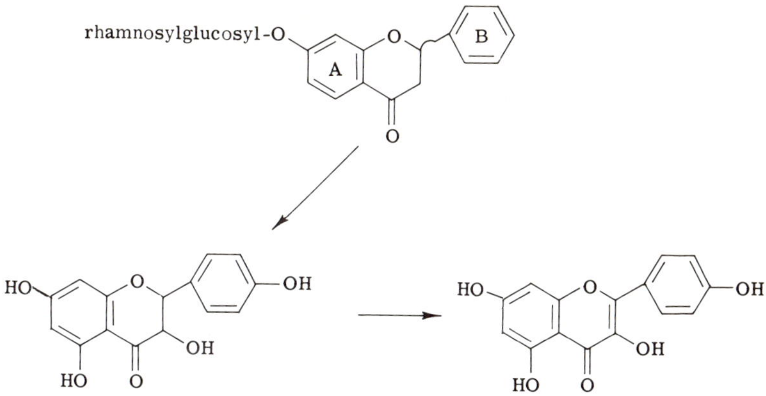

Dihydrokaempferol Kaempferol

Scheme 5.18. Breakdown of naringin to Kaempferol (Maier and Metzler, 1967a).

grapefruit juice and observed a substantial decrease in bitterness. An enzyme was isolated from the preparation, naringinase, which was active over the pH range 3–5. This enzyme was partially purified by Thomas *et al.* (1958), who found that it was composed of two glycosidases. Dunlap *et al.* (1962) separated these two enzymes and identified them as a rhamnosidase and a glucosidase. The reactions catalyzed by these enzymes are as follows:

$$\text{Naringin} \rightarrow \text{prunin} + \text{rhamnose} \xrightarrow{\beta\text{-glucosidase}} \text{naringenin} + \text{rhamnose} + \text{glucose}$$

These enzymes were reported to debitter grapefruit juice and pulp (Thomas *et al.*, 1958; Ting, 1958; Griffiths and Lime, 1959). Naringinase was later isolated from *Aspergillus niger* by selective destruction of pectin enzymes at pH 8.0 and 37°C for 2 hours (Omura *et al.*, 1963). This treatment did not effect the naringinase preparation and provides a commercial enzyme in Japan for removing naringin from the juice of bitter orange (*Citrus auranticum*). Goldstein and co-workers (1971) prepared water-insoluble derivatives of naringinase by linking the enzyme covalently to copolymers of maleic anhydride with ethylene, methyl vinyl ether, isobutyl vinyl ether, and styrene. These researchers claimed that an immobilized naringinase system could be used in a continuously generated bath or column process for removing the bitterness from clarified juices.

Chapter 6

Aldehydes, Alcohols, and Esters: Biogenesis

I. INTRODUCTION

The important flavor volatiles of fruits include short-chain unsaturated aldehydes and alcohols (C_3–C_6), ketones, and esters. Their biogenesis occurs during the short ripening period associated with the climacteric rise in respiration (Drawert *et al.*, 1966; Heinz *et al.*, 1965; Paillard, 1968; Romani and Ku, 1968; Tressl *et al.*, 1970). Although data on flavor biogenesis are rather scarce, nevertheless considerable research has been conducted on tomato and banana volatiles. This chapter discusses flavor biogenesis in these economically important fruits.

II. TOMATO VOLATILES

The typical aroma of field-ripened tomatoes is attributed to aldehydes and ketones (32%), short-chain alcohols (10%), and hydrocarbons, long-chain alcohols, and esters (58%) (Sha *et al.*, 1969). Many of these compounds have been identified, but their contribution to flavor remains incompletely understood (Kazeniac and Hall, 1970; Stone *et al.*, 1971, 1975; Sha *et al.*, 1969). Several alcohols were first isolated by Spencer and Stanley (1954) as 3,5-dinitrobenzoate derivatives from fractions eliciting tomato odor. Pyne and Wick (1965) identified *cis*-3-hexenol as a major volatile contributing to tomato aroma by its characteristic "green note." This alcohol was also detected by Johnson *et al.* (1968) among tomato volatiles and varied with the variety of the fruit and the stage during which it was harvested. Pyne and Wick (1965) also identified several aldehydes, including *n*-hexanal and *trans*-2-hexenal. These aldehydes were also observed by Dalal *et al.* (1967) among the 15 major volatile components of red ripe tomatoes. Dalal and co-workers (1968) investigated the volatiles produced during the ripening of tomatoes and noticed that with the exception of isovaleraldehyde and hexanol the levels of the components increased as the fruit matured. Isovaleraldehyde and hexanol concentrations were highest at the breaker and large green stages and appeared to contribute a "green" or "grassy" odor to the tomato. Field-grown tomatoes were richer in volatiles than greenhouse-grown tomatoes, as shown earlier by Dalal *et al.* (1967). This difference was considered responsible for the lack of bright red color or typical aroma associated with greenhouse-grown tomatoes. Sha *et al.* (1969) compared the volatiles of field-ripened, artificially ripened, and overripened tomato fruits but were unable to demonstrate clear differences since the levels of some components were higher in the field-ripened and others in the artificially ripened fruit. Viani *et al.* (1969) isolated 46 compounds from tomato volatiles including hexanal, *cis*-3-hexenal, *trans*-hexenal, *cis*-3-hexenol, hexanol, and 2-isobutylthiazole. The latter compound had a strong green leaf odor. Kazeniac and Hall (1970) identified 36 tomato volatiles, of which *cis*-3-hexenal and hexanal were found to be the major components of blended, crushed ripe tomatoes.

Johnson *et al.* (1968) monitored the levels of isoamyl and active amyl alcohols, *n*-pentanol, and *cis*-3-hexenol in two tomato varieties: Rutgers and KC 146. Both varieties were related genetically and exhibited similar changes with respect to these volatiles in the fresh juice obtained from different harvests. There was considerably more hexenol in the fruit obtained from the second harvest than in that obtained from the first harvest, although the amount present was still extremely small relative to that of isoamyl alcohol. In addition to these varieties, Johnson *et al.* (1968) also noted that Fireball, an early tomato variety, generally contained lower levels of solids as well as volatiles. The differences observed between harvest times were attributed to slight changes in weather and soil conditions. This could cause a corresponding change in composition as reflected by differences in volatiles. Stevens (1970) suggested that heritable differences could explain differences in levels of eugenol, 2-isobutylthiazole, and methyl salicylate in the volatiles from Campbell 146 and Campbell 1327 tomato cultivars. 2-Isobutylthiazole appeared to be determined by a single additive gene, and methyl salicylate and eugenol also were determined by single genes linked in coupling phase with a dominance for low concentration. Although 2-isobutylthiazole was present at much lower levels than either eugenol or methyl salicylate, its lower threshold indicated a greater contribution to tomato flavor.

A. Amino Acid Precursors

Considerable evidence pointed to the precursor role of certain amino acids in the formation of tomato volatiles. Dalal (1965) observed an increase in 3-methyl-1-butanol, isopentyl acetate, isopentyl butyrate, and isopentyl isovalerate levels during the ripening of tomatoes, whereas the concentration of 3-methyl-1-butanal increased up to the breaker stage and then gradually decreased. Yu *et al.* (1968a) noted that the alcohol portion of these esters had carbon skeletons similar to that of 3-methyl-1-butanal and suggested their synthesis from this aldehyde. The presence of this compound in tomato volatiles was reported by several researchers (Pyne and Wick, 1965; Schormuller and Grosch, 1964; Dalal *et al.*, 1967). The similarity between the carbon skeleton of leucine and 3-methyl-1-butanal led Yu *et al.* (1968a) to investigate the relationship between leucine and this aldehyde. They incubated L-[^{14}C]leucine with crude extracts of fresh tomatoes (*Lycopersicon esculentum* Mill), V. R. Moscow variety greenhouse-grown, and identified the products by gas chromatography. The results (Fig. 6.1) indicate three peaks in the presence of leucine two of which corresponding to 3-methyl-1-butanal and 3-methyl-1-butanol, respectively. A small single peak was evident in the control, 3-methyl-1-butanal, which was attributed to contamination of the enzyme. The inability of alanine or aspartic acid to produce 3-methyl-1-butanal when incubated with these preparations suggested the presence of a specific enzyme catalyzing the conversion of leucine to this aldehyde. The absence of any activity in the boiled extract in contrast to incorporation of L-[^{14}C]-leucine into labeled 3-methyl-1-butanol by the original extract confirmed the enzymatic nature of this reaction. Thus,

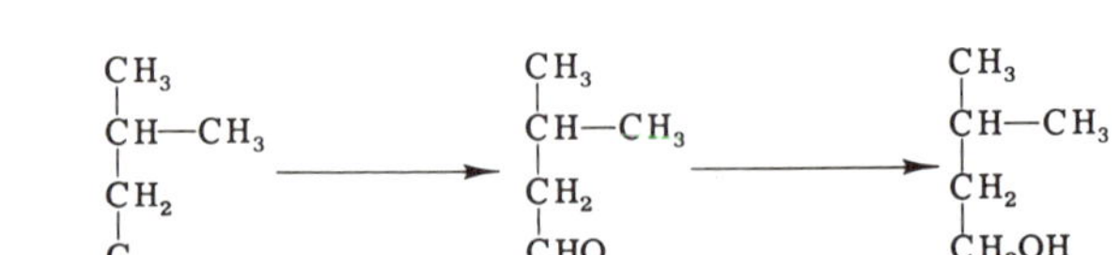

L-Leucine — 3-Methyl-1-butanal — 3-Methyl-1-butanol (isoamyl alcohol)

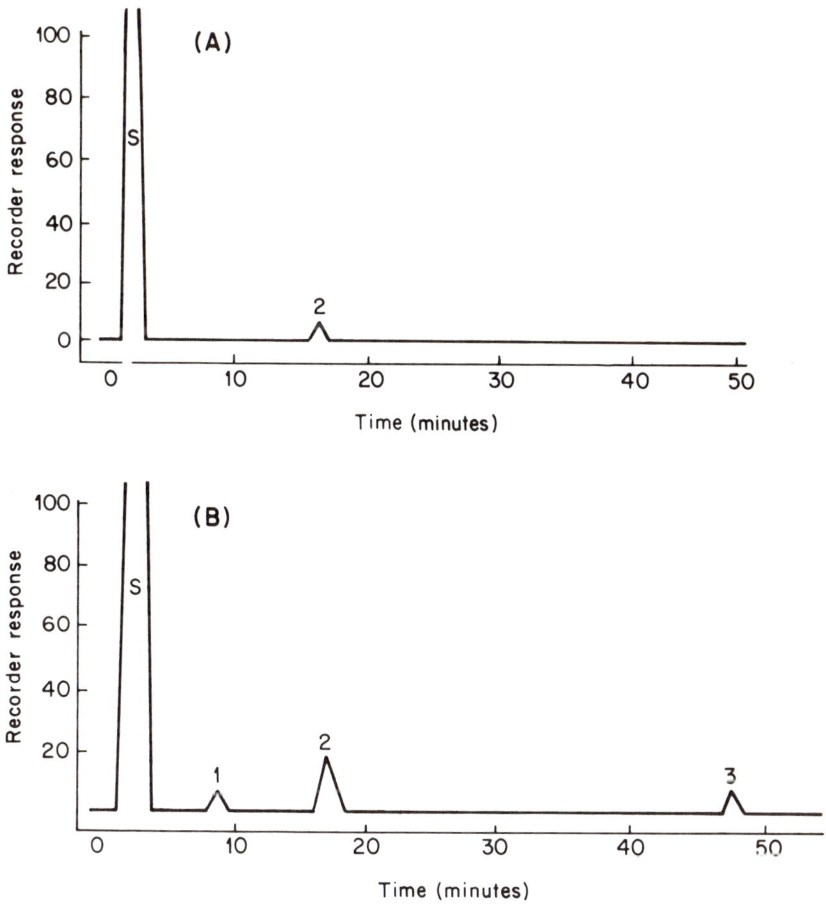

Fig. 6.1. Chromatograms of ether extracts of tomatoes from reaction mixtures with (B) and without (A) added leucine as substrate. Peak 1: 3-methyl-1-butanal; peak 2: 3-methyl-1-butanol; peak 3: unidentified [S, solvent (ether)]. (Yu *et al.*, 1968a. Copyright by Institute of Food Technologists.)

Further research by Yu *et al.* (1968b), however, demonstrated that, in addition to leucine, tomato extracts could also produce carbonyl compounds in the presence of alanine and aspartic acid. These amino acids were all present at high concentrations in the fruit, but these levels decreased during ripening. This suggested the presence of more intricate enzyme systems capable of utilizing a larger number of amino acids. Separation of the soluble and insoluble fractions by differential centrifugation of the tomato extracts showed that the enzymes involved were located at different sites. For instance, the soluble fraction converted leucine, whereas the mitochondrial fraction metabolized both alanine and aspartic acid. If the enzymes were transaminases these results would be consistent with reports by Wilson *et al.* (1954) and Bone and Fowden (1960) that alanine–glutamic acid and aspartic–glutamic acid transaminases were located in the mitochondria of plant cells. The presence of transaminases in tomato would explain the large amounts of glutamic acid reported by Freeman and Woodbridge (1960) and Yu *et al.* (1967). The activity of these enzymes was markedly higher in field-grown than in greenhouse-grown tomatoes, which probably explains the corresponding differences in carbonyl production (Dalal *et al.*, 1967). Yu *et al.* (1968c) monitored the production of glutamic acid when extracts of field-grown tomatoes at the green and red-ripe stages were incubated with alanine, aspartic acid, valine, and leucine. The results for tomato extracts at the green stage are summarized in Table 6.1. An increase in glutamic acid was evident in the presence of the amino acids, thus confirming the presence of transaminases. The reaction products varied with each tomato extract, although some differences were observed among substrates, incubation times, and extraction procedures. Propanal was detected in most reaction products, particularly in the presence of alanine. The extracts from red ripe tomatoes were more active than those obtained from the green fruit, producing primarily carbonyl compounds in the presence of alanine and alcohols with leucine and valine.

Yu and Spencer (1969) isolated α-ketoisocaproic acid among the products formed during the incubation of L-leucine with extracts of fresh

TABLE 6.1

Enzymatic Production of Glutamic Acid by Extracts of Tomatoes Harvested at the Large Green Stage[b]

	Substrate[b]			
	Alanine	Aspartic acid	Leucine	Valine
Glutamic acid formed (μmoles)	8.7[c]	10.6	6.9	9.6
Amino acid converted (%)	17.4	21.2	13.4	19.2

[a] From Yu *et al.* (1968c).

[b] Fifty micromoles each.

[c] Correction was made for glutamic acid originally present in the enzyme extracts. Glutamic acid was determined manometrically by the use of L-glutamic acid decarboxylase.

tomato fruit. This was confirmed with [^{14}C]leucine and established the presence of L-leucine : 2-ketoglutarate aminotransferase. The activity of this enzyme was located mainly in the supernatant fraction and decreased as the fruit matured. This was consistent with an earlier observation by Yu and co-workers (1968b). The activity of leucine aminotransferase in the mitochondria was less than 10% that in the soluble extract with no detectable activity in mitochondria from pink or red tomatoes. The reaction catalyzed by leucine aminotransferase is shown in Scheme 6.1, in which the formation of the corresponding alcohol is catalyzed by alcohol dehydrogenase.

Yu and Spencer (1970) also identified α-alanine : aminotransferase (EC 2.6.1.2) in tomato fruit. This was confirmed by isotopic studies with L-[U-^{14}C]alanine in which pyruvic acid was the most highly labeled product formed. The enzyme was active in both the supernatant and mitochondrial fractions, although Bone and Fowden (1960) found that it was predominant in the mitochondrial fraction. They suggested that α-alanine transferase may be prominent in the mitochondria since the pyruvate produced could be readily utilized in the Krebbs cycle. The activity of this enzyme declined in an analogous manner to that observed for leucine aminotransferase during ripening of the tomato fruit. This decline in activity was attributed in part to the presence of an inhibitor. This was based on work by Dickinson and Hanson (1965), who suggested that the decreased ability of tomato mitochondria to oxidize malate and pyruvate

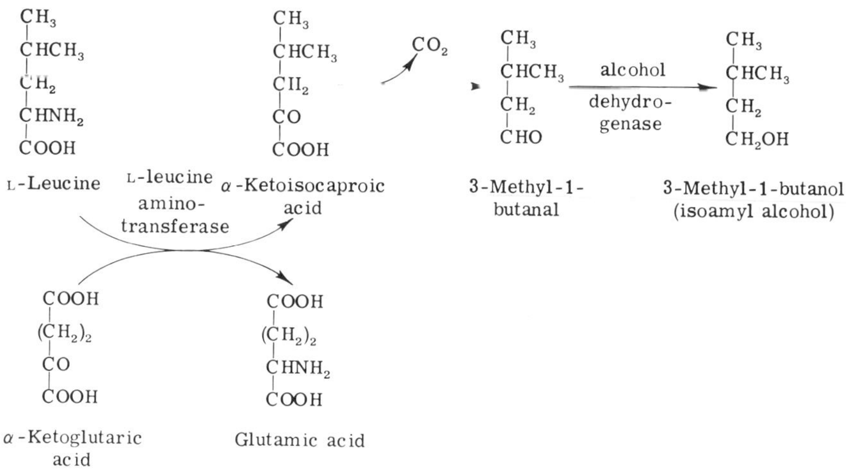

Scheme 6.1. Biogenesis of isoamyl alcohol from L-leucine in tomato fruit.

during ripening might be due to the appearance of an inhibitor. However, the nature of such an inhibitor remains to be characterized.

B. Fatty Acid Precursors

The aliphatic nature of the carbonyls of tomato volatiles suggested their synthesis from unsaturated fatty acids (Jadhav *et al.*, 1972). The major unsaturated fatty acids in the pericarp of tomatoes were reported by Kapp (1966) to be oleic, linoleic, and linolenic acids. Ueda and co-workers (1970) observed substantial amounts of lipids in green tomato fruits which decreased in fruits harvested at the breaker stage. As the fruit ripened there was a slight decrease in neutral lipids, which was accompanied by a drop in the levels of linoleic and oleic acids in the triglyceride fraction. Jadhav *et al.* (1972) also reported a corresponding decrease in linoleic and linolenic acids as the tomato fruit matured.

C. Lipoxygenase and Carbonyl Formation

Nye and Spoehr (1943) first noted the role of unsaturated fatty acids in volatile production. These workers observed the formation of hexenal in leaves of *Ailanthis glandulosa*, which occurred only in the presence of oxygen. The absence of hexenal in heat-treated samples implied the role of enzymes in its production. Later studies on the formation of α-hexenal in the leaves of *Ginkgo biloba* were reported by Major *et al.* (1963, 1968). Major and Thomas (1972) established linolenic acid as a precursor of hexenal in ginkgo leaves. Subsequent research indicated the importance of unsaturated fatty acids as precursors of short-chain aliphatic aldehydes in tomatoes (Galliard and Matthew, 1977; Jadhav *et al.*, 1972; Kazeniac and Hall, 1970; Stone *et al.*, 1975).

Kazeniac and Hall (1970) reported a twofold increase in the total amount of hexenals following the addition of linolenate to blended tomato extracts, which implicated this fatty acid as a precursor of the volatiles (Table 6.2). The absence of hexanal production in heat-treated (enzyme-inactivated) tomato extracts indicated the enzymatic nature of these reactions. These researchers suggested the role of lipoxygenase in the production of these volatiles when hydrogen peroxide, a known inactivator of this enzyme, inhibited the formation of hexenals.

Stone *et al.* (1975) verified the precursor role of linolenate when they injected [^{14}C]linolenic acid into tomatoes and detected incorporation of the label into *cis*-3-hexenal in the distillate of the blended fruit. Jadhav and co-workers (1972), using tissue slices or cell-free extracts of tomatoes,

TABLE 6.2

Effect of Linolenic Acid on Development of Aldehydes in Enzyme-Inactivated Blended Tomatoes[a]

System	Peak area (mm^2)			
	n-Hexanal	*cis*-3-Hexenal	*trans*-2-Hexenal	Total hexenals
Control (no heat)	141	38	15	53
Control (no heat) + 750 ppm linolenate	120	72	21	93
Heated sample	38	Trace	24	24
Heated sample + 750 ppm linolenate	40	Trace	29	29

[a] From Kazeniac and Hall (1970). Copyright by Institute of Food Technologists.

also reported incorporation of [^{14}C]linoleic and linolenic acids into *n*-hexanal. The inhibition of carbonyl formation and oxygen uptake by hydrogen peroxide in this study was consistent with the role of lipoxygenase. The participation of heme compounds in the degradation of unsaturated fatty acids was discounted since KCN did not adversely affect the production of *n*-hexanal. These workers observed an increase of lipoxygenase activity accompanied by a decrease in linoleic and linolenic acids during ripening of the tomato fruit. Lipoxygenase activity reached a peak at the onset of the climacteric period, which might explain in part the increase in volatile production reported by Dalal *et al.* (1968) during maturation of the tomato. Kazeniac and Hall (1970) also found volatile aldehydes such as *cis*-3-hexenal, *trans*-2-hexenal, and *n*-hexanol to be present at much higher levels in the fully ripened tomato than in the immature fruit. The formation of *trans*-2-hexenal was attributed by Kazeniac and Hall (1970) to the instability of *cis*-3-hexenal to the acidic pulp and juice as well as to temperature, resulting in isomerization to the *trans* form. This reaction was confirmed by the appearance of *trans*-2-hexenal in the distillate of processed tomato juice to which *cis*-3-hexenal was added. Thus,

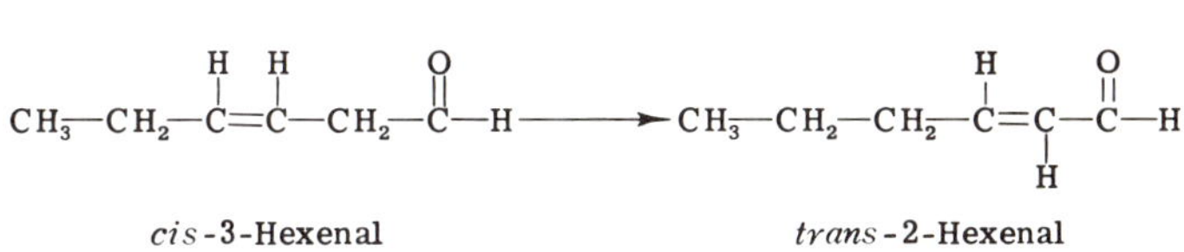

D. Sample Preparation and Volatile Composition

The preparation of tomato samples before analysis appeared to have a marked effect on the composition of the volatiles. Blended tomatoes were found to be higher in aldehyde content than was coarsely ground tissue. Kazeniac and Hall (1970) attributed this to the greater mixing of air facilitated by the blending procedure; however, the more thorough disintegration would permit the enzymes and substrates to interact more readily. This provided additional support for lipoxygenase activity, since atmospheric oxygen is introduced into the hydroperoxide molecule during the course of the reaction (Dolev *et al.*, 1967). Stone and co-workers (1975) also detected higher aldehyde levels in the volatiles of blended tomato preparations. They attributed this to the increased penetration of air facilitated by the extensive disruption of the cell. The major component of vacuum-distilled volatiles from blended tomatoes injected with ^{14}C-labeled linolenic acid was *cis*-3-hexenal; *cis*-3-hexenol and *n*-hexanol were the major components of steam-distilled volatiles of crushed tomatoes. When ^{14}C-labeled linoleic acid was injected, the major component of vacuum-distilled volatiles from the blended tomatoes was *n*-hexanal, whereas *n*-hexanol was the major component of steam-distilled volatiles of the crushed tomatoes.

Kazeniac and Hall (1970) observed an increase in *cis*-3-hexenol in the volatiles of blended tomatoes held for long periods of time, which suggested the presence of an oxidoreductase system. This was confirmed by Stone *et al.* (1975), who detected the rapid reduction of *cis*-3-[^{14}C]hexenal to *cis*-3-hexenol and hexanol in crushed tomatoes. A similar reduction was evident for *trans*-2-[^{14}C]hexenal to *trans*-2-hexenol and *n*-hexanol, although the rate of reaction was slower. This suggested that the aldehyde group was reduced more rapidly than the double bond, as shown previously in peas by Eriksson (1968). Stone and co-workers (1975) found that the reduction of *n*-[^{14}C]hexanal to *n*-hexanol approached 96% in crushed tomatoes. Schormuller and Grosch (1965) reported the presence of acetaldehyde, 3-methylbutanal, *n*-hexanal, and their corresponding alcohols in tomato volatiles and suggested that their interconversion involved an alcohol oxidoreductase. The enzymatic conversion of ethyl alcohol, isopropyl alcohol, and *n*-propanol to acetaldehyde, acetone, and propanal in tomato tissues was reported by Meigh *et al.* (1966) and Kazeniac and Hall (1970). Gremli and Wild (1976) confirmed this by the use of reconstituted spray-dried tomato powder.

In contrast to studies by Kazeniac and Hall (1970) and Stone *et al.* (1975), who showed *cis*-3-hexenal to be the major volatile of tomato distillate, Jadhav and co-workers (1972) reported *n*-hexanol to be the major volatile. Stone *et al.* (1975) attempted to explain this difference by suggesting that *cis*-3-hexenal acts as a precursor of *n*-hexanol. Only 2% of *cis*-3-[^{14}C]hexenal, however, was incorporated into *n*-hexanol, which appeared to eliminate it as an intermediate in hexanol formation. The differences observed were attributed by these researchers to the variability in the preparation of tomato samples. Kazeniac and Hall (1970) suggested that this also accounted for the inability of Nelson and Hoff (1969) to detect *cis*-3-hexenal, *trans*-2-hexenol, or *cis*-3-hexenol among the volatiles produced. Stone *et al.* (1975) proposed the pathway shown in Scheme 6.2 to explain the formation of these volatiles from ^{14}C-labeled linolenic and linoleic acids in tomatoes. This mechanism suggested that 13-hydroperoxide was the major product of the oxidation of linoleic and linolenic acids and was converted to *cis*-3-hexenal, *n*-hexanal, and *trans*-2-hexenal and their corresponding alcohols. In addition to the 13-hydroperoxide, Jadhav *et al.* (1972) proposed the formation of 9-, 12-, and 16-hydroperoxides of linolenic and linoleic acids to explain the presence of hexanol, propanol, 2,4-decadienal, 2,6-heptadiene, and *cis*-3-hexenal in tomatoes.

Galliard and Matthew (1977) were unable to demonstrate the formation of hexanol from linolenic acid as suggested by Jadhav *et al.* (1972). The major products formed when linoleic and linolenic acids were incubated with tomato homogenates were fatty acid hydroperoxides (50–80% yield). The major ones produced from linoleic acid were 9- and 13-hydroperoxides in the ratio 95 : 5. These researchers identified a hydroperoxide cleavage system that was active only toward the 13-hydroperoxide. One of the main nonvolatile products identified following incubation of linoleic acid 13-hydroperoxide with tomato extracts was 12-oxo-13-hydroxyoctadec-*trans*-1-enoic acid.

Further research by Galliard *et al.* (1977) on the enzymatic breakdown of lipids suggested the pathway shown in Scheme 6.3 to explain the formation of carbonyl fragments in disrupted tomato fruits. The phospholipids appeared to be degraded by phospholipase D (EC 3.1.4.4) due to the increased production of phosphatidic acid, although some hydrolysis was attributed to deacylation by acyl hydrolase enzymes. The hydrolysis of phospholipids as well as galactosyldiglyceride and triacylglycerols was thought to give rise to $C_{18:2}$ and $C_{18:3}$ fatty acids, which were subsequently oxidized to 9- and 13-hydroperoxides by lipoxygenase. Of these only the 13-hydroperoxide was cleaved by a hydroperoxide cleavage enzyme to form hexanal and *cis*-3-hexanal together with the nonvolatile 12-oxododec-*cis*-9-enoic acid. The predominance of *cis*-3-enals among the unsaturated

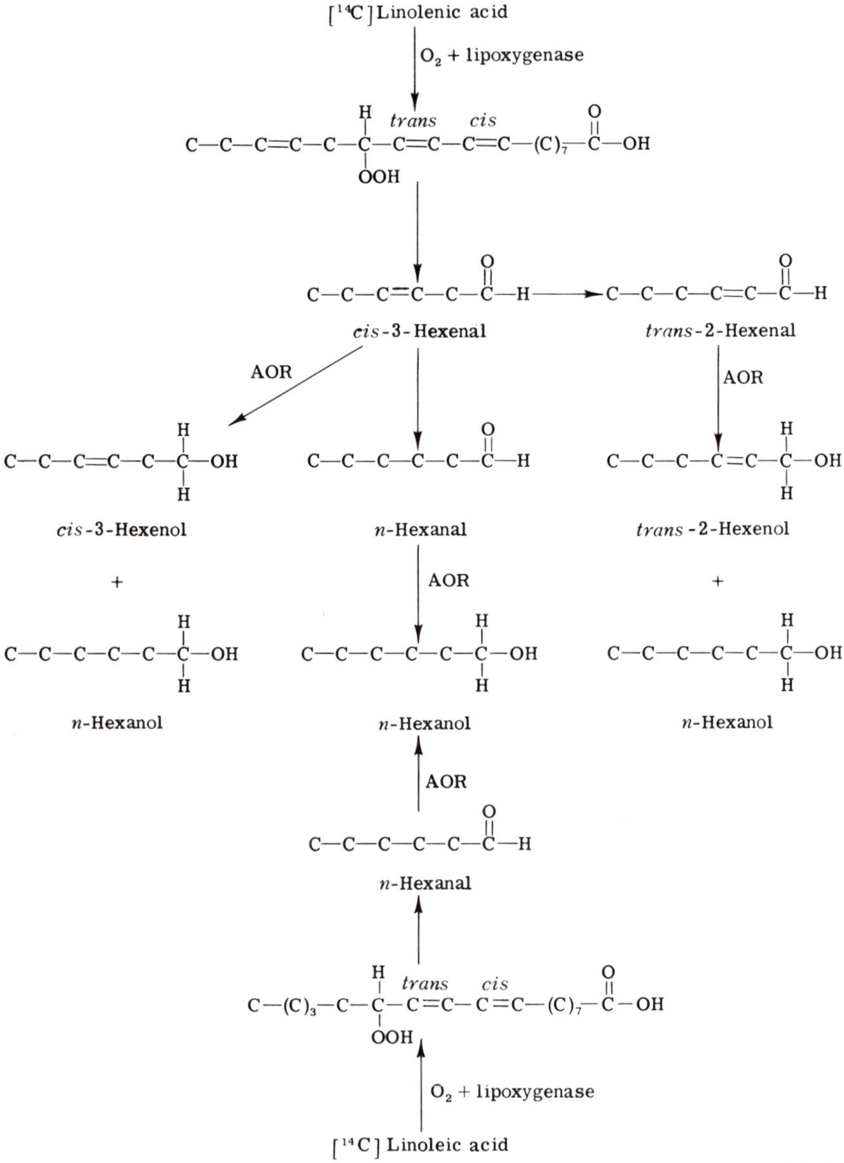

Scheme 6.2. Aldehyde and alcohol formation from linoleic acid in tomato (AOR, alcohol oxidoreductase) Stone *et al.*, 1975. Copyright by Institute of Food Technologists.

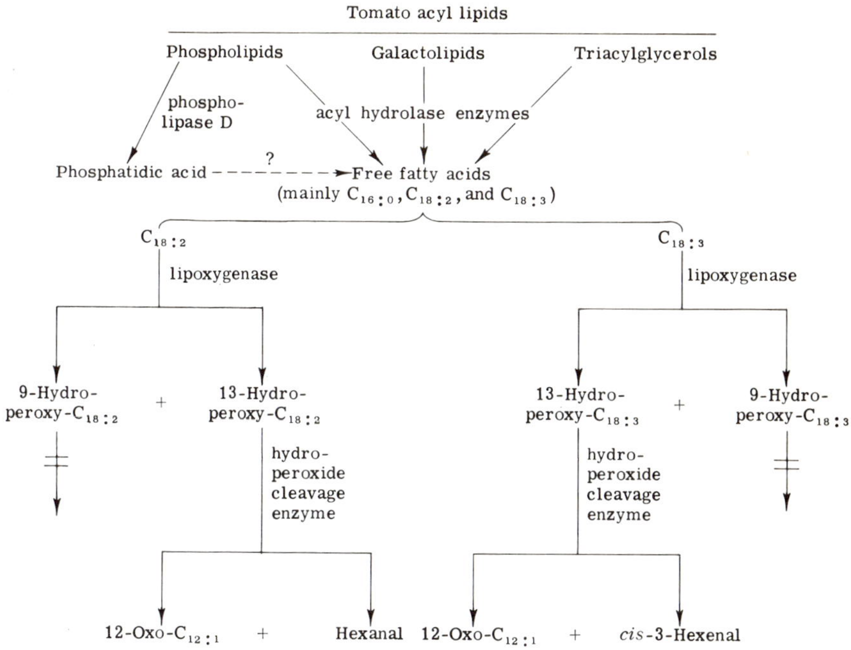

Scheme 6.3. Proposed pathway for the formation of carbonyl fragments by the enzymatic degradation of acyl lipids in disrupted tomato fruits (Galliard *et al.*, 1977).

aldehydes suggested the lack of isomerase enzymes in tomato extracts, in contrast to corresponding extracts from Ginko leaves (Major and Thomas, 1972) or cucumber (Hatanaka *et al.*, 1975).

Galliard *et al.* (1977) proposed the following mechanism to explain the cleavage of linoleic 13-hydroperoxide to hexanal and the C_{12} oxo acid:

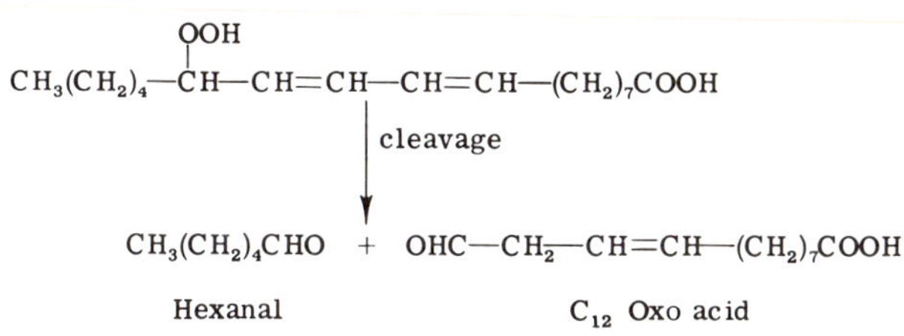

The major question as to the role of the 9-hydroperoxide remains to be resolved since it does not appear to be further metabolized.

III. SENSORY PROPERTIES OF SHORT-CHAIN ALDEHYDES AND ALCOHOLS

The importance of *cis*-3-hexenal in tomato aroma was attributed to its desirable blending and mouth-feel properties. This aldehyde was shown by Kazeniac and Hall (1970) to be effective in tomato juice over the range 0.3–0.5 ppm. Winter and Gautschi (1962) described the aroma of *cis*-3-hexenal as green fresh, which is characteristic of *cis* nonconjugated unsaturated compounds. The green odor of *cis*-3-hexenal was also recognized by Buttery *et al.* (1971), who emphasized its contribution to tomato aroma as a result of its low threshold in water (0.25 ppb). This contrasted with the decreased potency of *trans*-2-hexenal, which Buttery *et al.* (1971) reported to have a high threshold in water, 17 ppb. A greenish flavor was also detected following the addition of *n*-hexanal to tomato juice over a 0.0–0.5 ppm range; however, the potency was lower than that of *cis*-3-hexenal or *trans*-2-hexenal. Desirable mouth-feel and blending properties were also characteristic of conjugated dienals such as 2,4-decadienal.

Dalal *et al.* (1968) also attributed the green leafy aroma of tomato to *n*-hexanol and isovaleraldehyde. Kazeniac and Hall (1970) observed that three times the level normally found in tomato volatiles of *n*-hexanol were necessary to effect any flavor change. The importance of *cis*-3-hexanol to tomato aroma was recognized by Pyne and Wick (1965), although Kazeniac and Hall (1970) found that this compound lacked the flavor-enhancing properties and blending characteristics of aldehydes. The only component rated from moderately similar to similar to fresh tomato aroma was reported by Guadagni *et al.* (1972) to be *cis*-3-hexenal. The addition

TABLE 6.3

Effect of Enzyme Extracts of the Tomato Plant on Reconstituted Spray-Dried Tomato Juice[a]

Enzyme source	Flavor description
Stem	Green, earthy woody
Leaf	Green, sweet
Green fruit	Green, fresh
Yellow fruit	Fruity, slightly green
Ripe fruit	Fresh, slightly green

[a] From Gremli and Wild (1976).

of this compound (2 ppm), however, did not improve the flavor of tomato juice prepared by foam-mat-dried tomato powder, as established by a trained panel. Nevertheless, a mixture of *cis*-3-hexenal, 2-methylhept-2-en-6-one, eugenol, and β-ionone significantly improved the aroma of the foam-mat-dried tomato juice reconstituted product. The threshold values reported for these compounds were all established in water. However, although this provides useful information regarding the flavor potency of these compounds, it does not necessarily reflect the situation *in vivo*, in which synergistic or suppressive effects can occur.

IV. REGENERATION OF TOMATO VOLATILES

Gremli and Wild (1976) attempted to regenerate the volatiles of reconstituted spray-dried tomato powder in order to recover the lost flavor. Enzyme preparations were extracted from different parts of the tomato plant as well as from the fruit at different stages of maturity. These extracts were then evaluated for their ability to improve the flavor of the reconstituted product. The fullest flavor regeneration was associated with extracts from leaves or ripe fruit, as shown in Table 6.3.

The presence of lipoxygenase and transaminase in the ripe fruit appeared to be responsible for the production of the essential volatiles. The majority of the aldehydes had been reduced to the corresponding alcohols. The formation of these alcohols, which was mediated by alcohol dehydrogenase, was favored by the acidic pH of the tomato product and explained the increase in alcohols associated with holding crushed tomato juice for long periods of time. Eriksson (1968) demonstrated that the equilibrium constant of this reaction in peas favored alcohol formation at a low pH, which explained the alcohol content of the volatiles found with the reconstituted spray-dried tomato powder at pH 4.3.

V. BANANA VOLATILES

Bananas provide an excellent system for flavor study since they can be ripened under controlled conditions and the nature and quantities of the volatile components can be monitored with relative ease. Rothenback and Eberlein (1905) first reported the presence of amyl acetate, amyl valerate, and ethyl ester in the steam distillate of bananas. Kleber (1912) confirmed amyl acetate as a normal constituent of ripe bananas by steam distillation of the banana pulp, saponification, and oxidation of the distillate followed by analysis of the acid salt produced with silver nitrate. About 40 years later Von Loesecke (1950) identified amyl esters of acetic acid and butyric acid in addition to appreciable amounts of acetaldehyde in the pulp of ripe Lacatan bananas. Very small quantities of ethyl and methyl alcohols were also reported, although no experimental procedures were described in this study.

Hultin and Proctor (1961) attempted to determine the effect of ripening, processing, and storage on the volatile components of bananas. Since this work was carried out before the advent of gas chromatography, an attempt was made to use as mild a method as possible in order to minimize any deleterious changes in the volatile constituents. The volatiles were trapped from the pulp of peeled bananas by a low-temperature high-vacuum method of distillation, and various derivatives from the aqueous distillate were identified. These included acetic acid, methyl alcohol, ethyl alcohol, methyl acetate, ethyl acetate, 2-hexenal, 2-pentanone, and tentatively isoamyl alcohol, isoamyl acetate, and 2-octanone. The relative changes in these volatiles were followed in Gros Michel bananas during four stages of ripening, as shown in Table 6.4. As the bananas ripened, the changes in color from a green skin color to full yellow to full yellow flecked with brown were accompanied by the production of a delicate aroma absent in the unripe fruit. 2-Hexenal appeared to make an important contribution to odor and, although not necessarily present in the fruit, was apparently produced on exposure of the tissue to air. The level of esters in the ripe stage was extremely low, with ethyl acetate being absent, although it is the predominant ester during the other stages of ripening. Although the majority of the compounds identified did not contribute directly to the "typical" banana odor, they were nevertheless important in providing the background notes. Exceptions to this, however, were 2-hexenal and isoamyl alcohol, both of which contributed to the "rank" odor of overmature bananas.

A comparison of the volatiles from heat-processed puréed bananas of Cros Michel and Lacatan origin indicated few differences between the two purées, although marked differences were observed when the fresh fruits were compared. The complete absence of 2-hexenal in canned banana purée suggested the participation of oxidative enzymes (Nye and Spoehr, 1943) in addition to the decomposition of 2-hexenal during processing and storage. The rather low ester content of the purées suggested that considerable hydrolysis occurred at elevated temperatures. This was suppported by the levels of acetic acid, which were higher than those in the fresh fruit.

TABLE 6.4

Some Volatile Constituents of Gros Michel Banana Pulp of Different Ripeness[a]

	Stage			
Compound[b]	Green	Half-ripe	Ripe[c]	Overripe
Acetic acid	0.26	0.56	0.48	1.57
Methyl alcohol	2.2	18.5	1.5	5.6
Ethyl alcohol	1.1	0.4	0.5	22.6
Isoamyl alcohol[d]	0.3	0.4	0.1	5.6
Methyl acetate	4.5	9.5	1.3	5.0
Ethyl acetate	22.4	9.5	0	16.9
Isoamyl acetate[d]	0.02	0.1	0.02	1.2
2-Hexenal	2.2	4.6	7.6	1.9
2-Pentanone	2.1	1.4	2.7	1.1
2-Octanone	0.3	0.2	0.8	0.8

[a] From Hultin and Proctor (1961). Copyright by Institute of Food Technologists.

[b] Expressed as milligrams per 100 gm pulp.

[c] In the ripe sample, a fourth carbonyl compound was observed. It was present in trace quantities, and no identification was made, but it was judged, by its R_f values to possibl contain about 10 carbon atoms.

[d] Tentative identification.

Hultin and Proctor (1962) attempted to regenerate the fresh banana odor in heat-processed banana purée using a crude enzyme preparation. This was based on the flavor precursor concept, in which the more stable flavor precursor survived the heat treatment although the enzymes and flavor volatiles were destroyed or lost. Besides preparing purée by high-temperature short-time processing of fully ripe bananas, these authors examined a variety of possible precursors for their ability to restore the original odor. Organoleptic evaluation indicated that much of the "fresh" banana odor could be restored to heat-processed banana purée by addition of the crude enzyme extract. The absence of odor regeneration in the presence of heated enzyme (80°C for 15 minutes) confirmed the enzymatic nature of the process. The most active preparations were obtained from the pulp of bananas ranging in skin color from half-yellow and half-green with green tips to full yellow. Among the compounds tested, pyruvic acid, valine, and oleic acid proved to be the most effective precursors under the conditions of the study and increased the rate of odor regeneration on standing. Leucine was not found to increase the rate of "fresh" banana odor even though it is known to be a precursor of isoamyl alcohol in bananas (Hultin, 1959), which is responsible for the rank odor of the overmature fruit.

A. Gas Chromatographic Patterns and Flavor Profiles

McCarthy *et al.* (1963) attempted to correlate the gas chromatographic patterns of banana volatiles with the flavor profiles of the fruit over successive stages of ripening under controlled conditions. The headspace above the fruit was examined by gas chromatography, and the flavor profiles of two varieties of yellow banana were compared. The Gros Michel exhibited a low flavor amplitude, with various characteristic notes ranging from threshold to slight to moderate. The typical Gros Michel banana flavor was described as "estery," with amyl acetate the only recognizable ester. In contrast, the Valery banana had a moderate to high flavor amplitude, with more complex "estery" notes described as being of moderate to strong intensity. The taste panel attributed the "estery" flavor to a combination of moderate amounts of amyl acetate together with threshold quantities of amyl propionate and amyl butyrate. They also attributed the slight to moderate "fruity" note to the butyl ester of butyric acid or valeric acid. Several other esters, such as butyl and hexyl acetates and butyl butyrate, provided a fruit impression. A fuller and more interesting flavor was associated with the Valery fruit. Confirmation by gas chromatographic studies showed fewer compounds and less volatile components for the Gros Michel compared to the Valery fruit (Fig. 6.2). Further examination of these volatiles for their predominant flavor characteristics resulted in three general sensory impressions: bananalike, fruity, and green and woody or musty. The compounds identified for Gros Michel had a bananalike flavor but little or no fruitiness, whereas Valery fruit had considerable amounts of compounds with a bananalike odor and fruity sensations. Several alcohols and carbonyl compounds were also found to possess green and woody flavors, which may contribute in part to these notes. The panel reported the green note to be similar to that of *trans*-2-hexenal, reported in banana by Hultin and Proctor (1961) and Issenberg and Wick (1963), which was also found in some chromatograms in this study.

Chromatographic patterns of the volatiles produced at successive stages of ripening in the Gros Michel and Valery fruits are shown in Fig. 6.2. In the green preclimacteric fruit, few or no volatiles were evident; however, following ripening the two varieties appeared to fall into two distinct groups. In the Valery fruit, the production of volatiles commenced during the climacteric phase and increased in amount and complexity within 12–24 hours after the climacteric peak, when the peel was yellow-green. A gradual increase in volatile content was evident during the postclimacteric period, when the fruit became fully yellow and finally flecked with brown. In contrast, the Gros Michel variety did not exhibit any significant production of volatiles until the late postclimacteric period, when the peel was completely yellow. This fruit was associated with fewer volatiles and less total volatile material at all stages of development. This was evident during the early stages of development from the very simple gas chromatographic patterns as well as flavor profile. Individual esters were present in low concentrations, and the flavors were described as a "general banana" or "estery." The increase in ester level resulted in the recognition of specific esters and recognition of individual notes. The intensity of these notes correlated with relative concentrations of the esters in the fruit vapor. Green and woody notes were reduced or absent in the ripe fruit even though chromatographic analysis still showed the presence of compounds known to produce these notes. It was apparent, therefore, that the more concentrated esters probably masked these compounds. Issenberg and Wick (1963) examined the flavor profiles of bananas by gas chromatography using direct vapor analysis and identified 2-pentanone, *n*-butyl acetate, isobutanol, *n*-butanol, *trans*-2-hexenal, and isoamyl butyrate in banana concentrates. Isobutyl acetate, isobutanol, and isoamyl butyrate had not been reported previously. Several other compounds were tentatively identified including 2-pentanol, *n*-propanol, *n*-amyl acetate, *n*-hexyl acetate, and methyl acetate.

B. Amino Acid Precursors

The importance of amino acids as precursors of banana flavor was demonstrated by Myers *et al.* (1969) using direct gas chromatographic vapor analysis. They observed the *in vitro* production of isoamyl acetate and isoamyl alcohol in bananas at different stages of maturity and examined the possible role of L-leucine as precursor. The production of isoamyl acetate was demonstrated in unripe slices in the presence of metabolic inhibitors but could not be shown in ripe or overripe slices since the amount of isoamyl acetate in the headspace remained essentially constant. The vapor concentration of isoamyl alcohol also appeared to be unchanged throughout all ripening stages. Whether these compounds were produced by the slices or utilized in some way as fast as they were produced was not

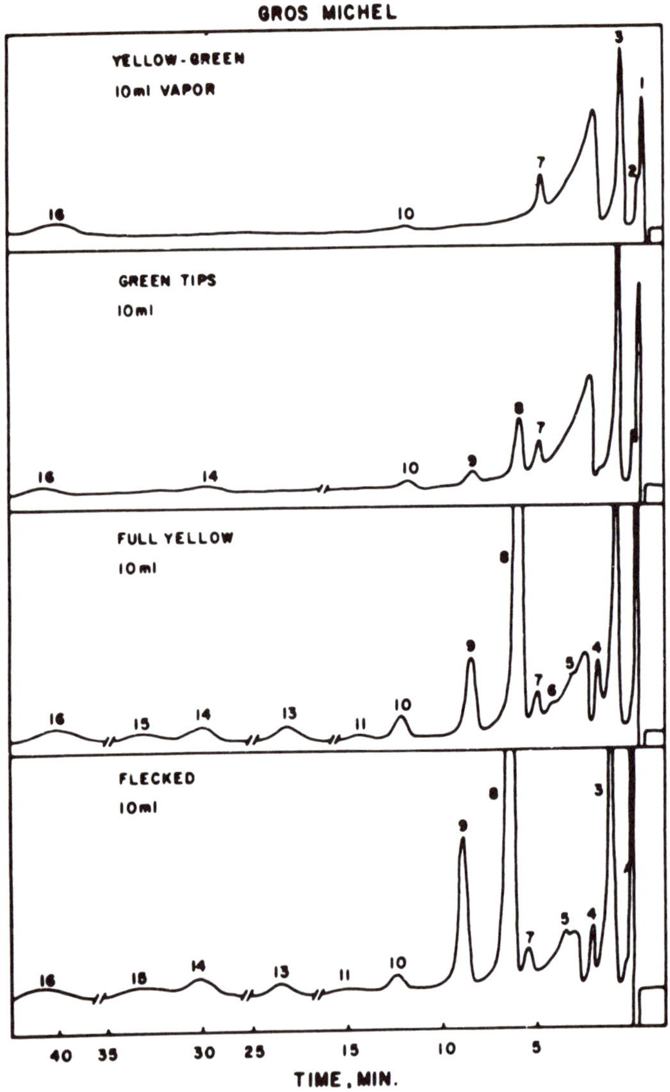

Fig. 6.2. Vapor chromatograms of ripening banana fruit varieties (tailing peak following

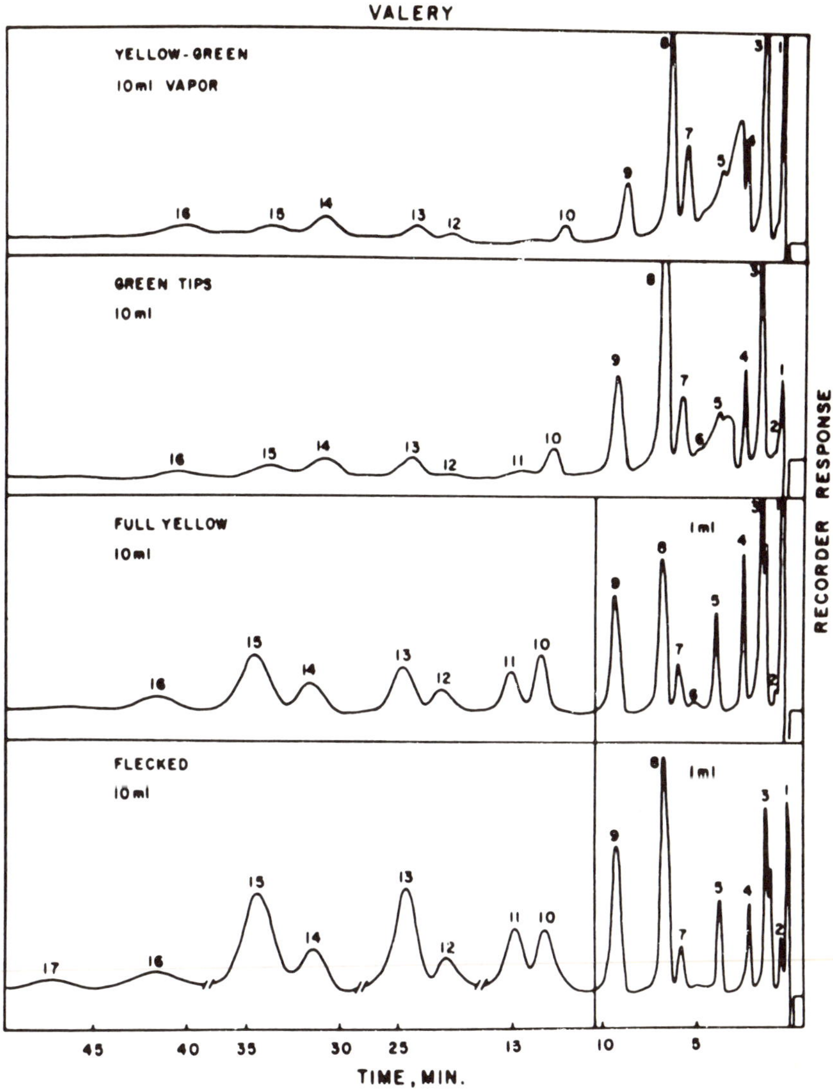

no. 4 is water vapor) (McCarthy *et al.*, 1963).

clear. The addition of leucine had no effect on the levels of isoamyl acetate or alcohol in the slices and appeared to rule out leucine as a possible precursor. However, when L-[U-^{14}C]leucine was used, the specific activities of both isoamyl acetate and isoamyl alcohol showed conclusively that leucine was a precursor of both these compounds.

The misleading interpretation arising from the direct vapor analyses was that the constant vapor composition provided no evidence that components were not being synthesized. Myers *et al.* (1970) confirmed that L-leucine is a precursor of isoamyl alcohol and isoamyl acetate, both volatile aroma constituents of banana fruit disks. L-Leucine was examined since it was reported to accumulate in ripening bananas of Gros Michel and Valery varieties (Steward *et al.*, 1960; Wick *et al.*, 1969), and a correlation between the increase in L-leucine and flavor production was observed. The conversion of branched-chain amino acids to branched-chain alcohols by *Saccharomyces cerevisiae* observed by Guymon (1965) pointed to a similar situation in bananas, in which branched-chain alcohols and esters are among the major aroma constituents. A study of the relationship between L-[U-^{14}C]leucine and isoamyl alcohol and acetate during several stages of ripening indicated that the major proportion of radioactivity resided in isoamyl alcohol and that only a small amount was incorporated into isoamyl acetate. When the incubation periods were increased, the proportion of radioactivity in both isoamyl alcohol and isoamyl acetate increased so that after 5 hours they corresponded to 78 and 10%, respectively (Table 6.5). Hydrolysis of isoamyl acetate indicated that part of the radioactivity was derived from acetate. Myers *et al.* (1970) pointed out, however, that under the conditions of their experiment, in which the slices were continually immersed in aqueous solution, unnatural behavior could occur (Beileski and Laties, 1963; Laties, 1964).

Tressl and Jennings (1972) developed an enrichment technique to investigate the continuous production of volatiles and observed their production during the postclimacteric phase. The two main classes of esters in ripe bananas were acetate and butyrate esters, although some 3-methyl butyrates, 2-methyl propionates, and caproates were also present in small amounts. Additional volatiles included some primary and secondary alcohols, ketones, and aldehydes in addition to phenolic esters. Although the butyl derivatives appeared to be derived from leucine, the corresponding 2-methyl propyl esters, 2-methyl propionates, and 2-methylpropanol could be theoretically derived from valine, since valine as well as leucine accumulates in ripening bananas (Palmer, 1971; Wick *et al.*, 1969). Tressl and Drawert (1973) found that L-leucine and L-valine in-

TABLE 6.5

Time Course of the Distribution of Radioactivity among Volatile Constituents from Ripe Banana Disks[a]

Constituent	Percent total radioactivity in volatiles[c]							
	10 min	20 min	30 min	1 hr	2 hr	3 hr	4 hr	5 hr
Peaks eluted before isoamyl acetate	14.5	22.6	17.0	6.9	4.0	2.6	1.4	1.9
Isoamyl acetate	1.5	2.2	2.3	3.6	5.7	7.6	7.0	10.0
Isoamyl alcohol	57.6	60.0	64.7	77.7	81.0	78.3	76.4	78.3
Peaks eluted after isoamyl alcohol	26.4	15.2	16.0	11.8	9.3	11.5	15.2	9.8
Total dpm in traps	59,555	76,655	97,872	166,286	233,543	286,250	311,976	565,000

[a] From Myers *et al.* (1970).

[b] Conditions: 12 slices incubated in 14.5 ml of 0.05 *M* phosphate buffer plus 0.5 ml L-[U-^{14}C]leucine (50 μCi 0.2 μmole).

[c] Percent total radioactivity is equal to $\frac{\text{dpm in a given trap}}{\text{total dpm in traps}} \times 100$.

creased from 5 to 15 mg per 100 gm in ripening bananas, whereas the amount of L-isoleucine and other amino acids remained constant. They showed that ^{14}C-labeled leucine was converted to 3-methyl-1-butanol, 3-methyl butyl esters, 3-methyl butyrates, and 2-ketoisocaproate. Hydrolysis of the aroma extract showed equal distribution of radioactivity in alcohols and acids. The largest proportion of radioactivity was observed in 3-methylbutanol, whereas 2-ketoisocaproic acid and 3-methylbutyric acid accounted for 97% of the radioactivity in the acid fraction. Experiments with ^{14}C-labeled valine in postclimacteric banana tissue slices showed an analogous picture, in which the label was observed in 2-methyl-1-propanol (4.5%), 2-methylpropionic acid (45%), 2-ketoisovaleric acid (22%), and 2 methyl propyl acetate (14%).

Gilliver and Nursten (1976) investigated the source of the acyl moiety in volatile ester formation in bananas; C_4 and C_2 acyl derivatives were incubated with banana slices and the formation of butyrates and acetates monitored. A fivefold increase in the headspace of isobutyl and isopentyl butyrates was observed following the addition of butyryl-CoA compared to only marginal increases in acetic acid esters in the presence of acetyl-CoA. Incubation of banana slices with the appropriate alcohol and the acyl source resulted in a further increase in headspace ester concentrations. This was evident for isobutyl and butyl butyrates, the levels of which increased 70-fold. These results indicated the ability of acyl-CoA to act as acylating agent in the formation of esters. Nevertheless, Gilliver and Nursten (1976) did not rule out the possible role of acyl carrier protein (ACP) derivatives in esterification since ACP can form biochemically from the CoA compounds. The esterification process appeared to involve essential factor(s) from banana slices, but whether these are enzymes or cofactors remains to be ascertained.

C. Fatty Acid Precursors

Drawert *et al.* (1966) reported the formation of hexen-2-al and *n*-hexanal following maceration of leaves and fruits. The production of these covolatiles appeared to be common in grape, banana, pear, apple, and plum homogenates. The inability of KCN to inhibit the production of these aldehydes suggested their genesis from linoleic and linolenic acids by the action of lipoxygenase. Bananas were associated with considerable amounts of these volatiles and have been the subject of further study.

The lipid content of bananas was reported to be 0.2–0.5% of the fresh weight of the pulp and remained fairly constant during storage. The major fatty acids of pulp and peel are palmitic, oleic, and linoleic acids. The amount of linoleic acid observed by Goldstein and Wick (1969) in ripe banana pulp was three times less than that in the unripe fruit, which suggested a possible relationship between the level of this fatty acid and the production of volatiles. Hultin and Proctor (1961) found 2-hexenal to be the major carbonyl in banana volatiles, reaching a maximum in the ripe (all yellow) fruit. These workers (Hultin and Proctor, 1962) attempted to regenerate the flavor of heat-processed banana pulp purée containing linolenic acid by the addition of crude extract from fresh banana pulp purée. An undesirable, strong fish odor, that was not evident in the control (without crude enzyme) emanated from the enzyme-treated purée. This was attributed to the presence of oxidizing enzymes in the crude extract, which may not be the case *in vivo*, where protection is afforded by the oxidation-reduction system of the plant. Issenberg and Wick (1963) reported the presence of *trans*-2-hexenal in the volatiles of Gros Michel bananas, and Murray *et al.* (1968) identified the presence of hexenol, *cis*-3-hexenol, and *trans*-3-hexenol among the volatile alcohols of ripe Cavendish bananas. Although *cis*-3-hexenol was present in very small amounts, it is recognized as a major volatile of many plants. Several unsaturated aldehydes were also observed including *trans*-4-hexenal and *cis*-4-hexenal. In an investigation of the neutral and acid fractions obtained by hydrolysis of odor concentrates from fresh banana pulp, Wick and coworkers (1969) failed to detect 3-hexenols or 2-hexenols. Using an enrichment technique, Tressl and Jennings (1972) analyzed the headspace over ripening bananas and reported the presence of *trans*-2-hexenal. Palmer (1973) separated the volatile fractions of bananas and identified *cis*-2-hexenal, *trans*-2-hexenal, and *n*-hexanal among the aldehydes.

The relationship between maturity of the banana fruit and aldehyde formation was studied by Tressl and Drawert (1973). Green banana homogenates were found to produce *trans*-2-nonenal, *trans*-2,*cis*-6-nonadienal, and 9-oxanonanoic acid. These compounds were identical with those reported for cucumbers by Fleming *et al.* (1968). Tressl and Drawert (1973) treated green bananas with ethylene and then stored them for 4 days at 15°C. This resulted in the production of hexanal, *trans*-2-hexanal, and 12-oxo-*trans*-10-dodecenoic acid, all of which were important contributors to banana flavor as a result of their low threshold values. Incubation of [^{14}C]linoleic and [^{14}C]linolenic acids with banana homogenates resulted in the incorporation of ^{14}C into these volatile components. In green bananas, however, labeled linoleate was incorporated into 9-oxanonanoic acid. The enzymes responsible for the formation of these aldehydes were difficult to isolate, and a purified extract of lipoxygenase failed to form any of these C_6 or C_9 aldehydes. However, a crude extract from ripe bananas transformed labeled linoleate into hexanal, hexanol, and 12-oxo-*trans*-10-dodecenoic acid. Since the primary products of lipoxygenase are hydroperoxides, Tressl and Drawert (1973) incubated radiolabeled 13- and 9-hydroperoxyoctadecadienoic acids with a crude banana enzyme extract. These compounds were subsequently converted to C_6 and C_9 aldehydes and the corresponding oxo acids.

On the basis of these results Tressl and Drawert (1973) proposed the mechanism shown in Scheme 6.4 to explain the formation of hydroperoxy acids by lipoxygenase and their subsequent breakdown to aldehydes and oxo acids by a second enzyme referred to as "aldehyde lyase." Vick and Zimmerman (1976) reported the presence of an enzyme referred to as "aldehyde lyase" in germinating watermelon seedlings. This enzyme ap-

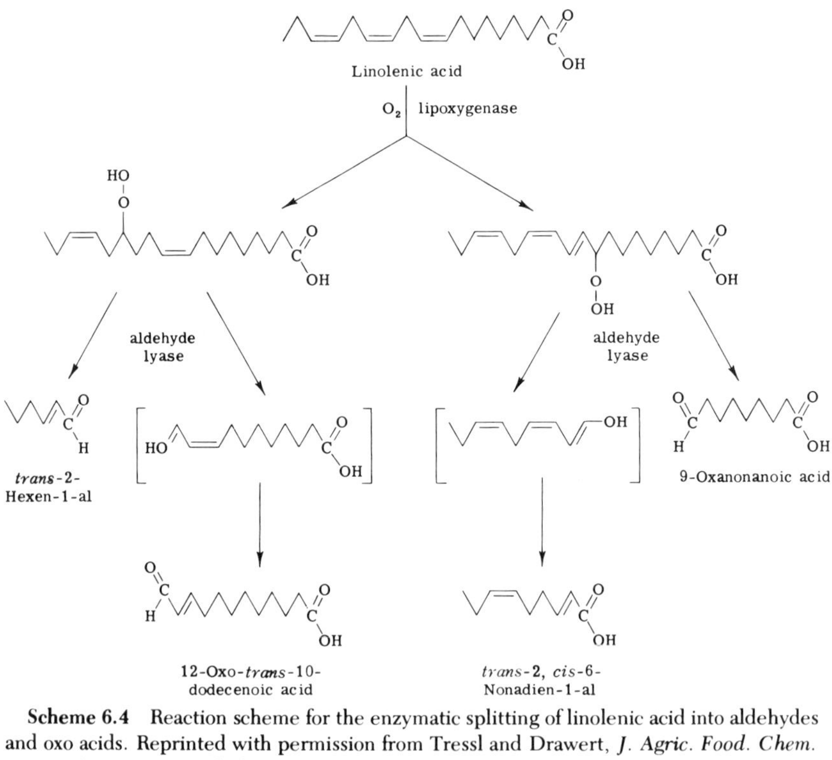

Scheme 6.4 Reaction scheme for the enzymatic splitting of linolenic acid into aldehydes and oxo acids. Reprinted with permission from Tressl and Drawert, *J. Agric. Food. Chem.* **21,** 560 (1973). Copyright by the American Chemical Society.

peared to be similar to the "aldehyde lyase" proposed by Tressl and Drawert (1973) and catalyzed the formation of 12-oxo-*trans*-10-dodecenoic acid and hexanal from 13-hydroperoxy-*cis*-9,*trans*-11-octadecanoic acid. Hydroperoxide lyase was not found in barley, corn, flax, green beans, peas, pumpkins, squash, or sunflowers. Unlike the hydroperoxide cleavage enzyme from tomatoes, which forms *cis*-3-enals as products (Scheme 6.3), hydroperoxide lyase appears to form *trans*-2-enals as primary products (Scheme 6.4).

SECTION III

Textural Components of Food: Chemistry, Physical Structure, and Effects on Quality

The texture of edible plant parts is determined by the structure and organization of the cell walls and intercellular cementing substances, which together form the plant tissues. The chemical and biochemical reactions taking place within the cell walls provide insight into the physical properties of fruits and vegetables that change during development, storage, and processing. A study conducted by Szczesniak and Kleyn (1963) on consumer awareness of texture showed that, for such raw foods as celery, apples, lettuce, watermelon, and pears, texture was far more important than flavor. This section discusses the various components of the cell wall in relation to the texture of plant foods.

Chapter 7

The Plant Cell Wall: Cellulose and Hemicellulose

I. INTRODUCTION

The cell wall is composed of three distinct parts: an intercellular cement or middle lamella, a primary cell wall, and a secondary cell wall (Fig. 7.1). The middle lamella, jellylike in consistency, forms between

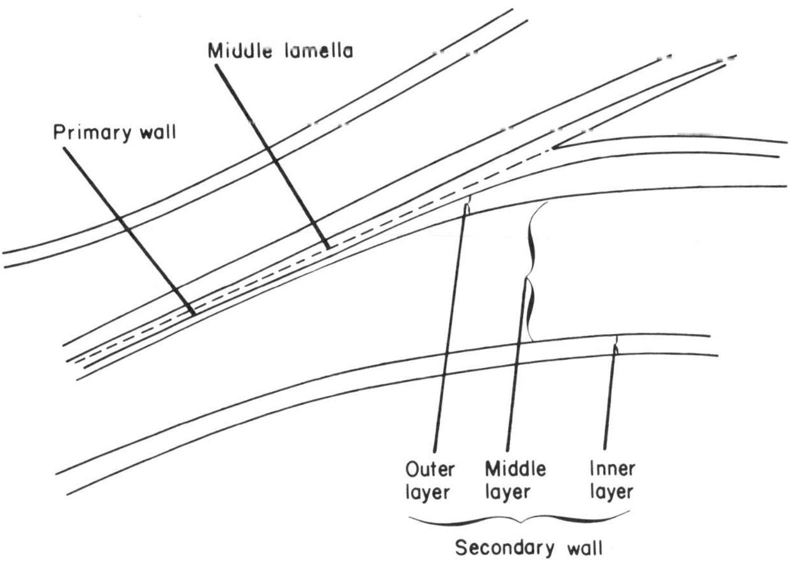

Fig. 7.1. Components of the plant cell wall.

adjacent cells during the course of cell division. It is extremely rich in pectic substances, which bind the plant cells together like cement (McClendon, 1964). The primary cell wall, located between the plasma membrane and middle lamella, is formed quite early in the development of the cell. A thin, flexible layer, it forms the outer covering of the cells of fruits. roots, and fleshy stems. Polysaccharides are the major group of compounds in the plant cell walls and are traditionally grouped into cellulose, pectic substances, and hemicelluloses (Northcote, 1963). These compounds are classified according to their solubilities rather than their chemical composition. Pectic substances are the fraction extracted with cold and hot water, hemicelluloses are extracted with alkaline solutions,

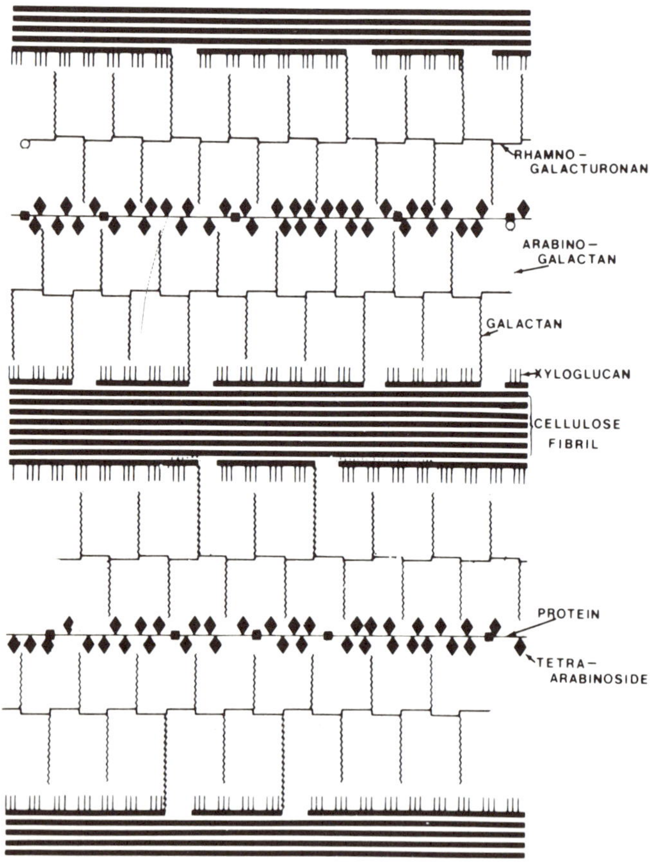

Fig. 7.2. Proposed model for the primary plant cell wall (Keegstra *et al.*, 1973).

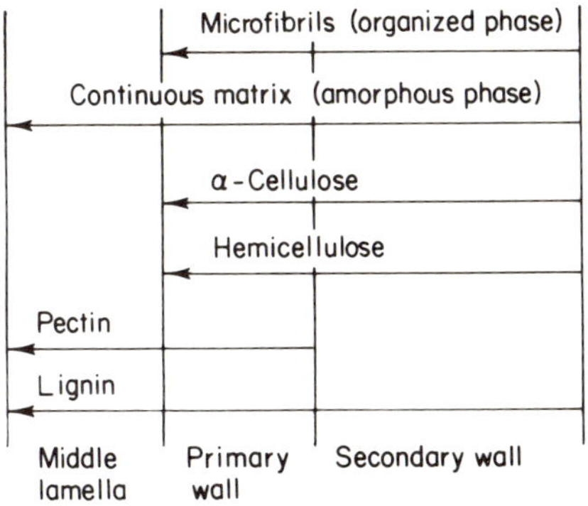

Fig. 7.3. Distribution of some of the components in the cell wall. Reprinted with permission from Worth *Chem. Rev.* 67, 465 (1967). Copyright by the American Chemical Society.

and cellulose is the residue, which is unaffected by either of these extraction procedures. In addition to these polysaccharides, the presence of a protein fraction in the primary cell wall has also been established (Lamport, 1965). Keegstra *et al.* (1973) proposed a model for the primary cell wall (Fig. 7.2) in which protein is covalently linked to the pectic substances through arabinogalactan.

When growth ceases certain mature cells form a secondary cell wall, which provides the plant cell with strength and mechanical support. It is during this period that lignin is synthesized and deposited in the cell walls. Figure 7.3 outlines the distribution of the various chemical components of the cell wall. In this model, the cell wall is an organized phase of cellulose microfibrils in a continuous matrix of noncellulosic material (Worth, 1967).

II. CELLULOSE MICROFIBRILS

The main feature of the cell walls of all green plants is a framework of organized aggregates of cellulose molecules, referred to as microfibrils, embedded in a matrix of noncellulosic polysaccharides and lignin. Cellulose is composed of long chains of β-1,4-linked D-glucopyranose residues containing the repeating unit of cellobiose. The individual chains of glucan are held together in fibrils by hydrogen bonds between the hydroxyl groups at carbon 6 and the glycosidic oxygens of the adjacent chains (Northcote, 1972). The organization of the cellulose chains within the microfibrils has been the subject of considerable controversy. Many models have been proposed and are discussed in several comprehensive reviews (Northcote, 1969; Ranby, 1969; Shafizadeh and McGinnis, 1971). A brief outline of some of the models is included here.

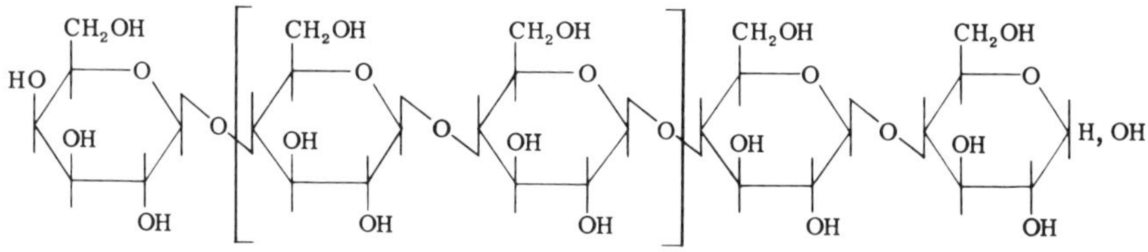

Cellobiose

The first generally accepted model was based on the fringe-micellar theory of Kratky (1935) and Frey-Wyssling (1936), who suggested that microfibrils were 15–20 nm in diameter and consisted of about 600 cellulose molecules. These microfibrils were composed of crystalline and amorphous regions. The crystalline regions were composed of rod-shaped, highly organized micelles 5–6 nm in diameter and produced discrete X-ray patterns. The background scattering was attributed to the less orderly arrangement of the cellulose molecules in the amorphous phase. A later model proposed by Preston and Cronshaw (1958) described the microfibril as a crystalline core of highly organized cellulose molecules surrounded by a paracrystalline sheath 50–100 Å in cross section (Fig. 7.4a). In a model proposed by Hess *et al.* (1957) the microfibril was composed of elementary fibrils, each containing 15–40 cellulose molecules and having a cross-sectional dimension of around 3.5 nm. The cellulose molecules were less organized at certain locations along the microfibril (Fig. 7.4b). This model differed from the fringe-micellular model in the regular arrangement of the crystalline and amorphous regions.

The concept of crystalline and amorphous regions in the microfibrils was disproved by Mühlethaler (1960), who examined these regions in ramie fibers and other cellulosic materials by electron microscopy after staining with phosphotungstic acid. Normally, this acid would penetrate the capillary pores and loose textures associated with amorphous regions, giving a dark contrast when examined under the electron microscope. However, Mühlethaler (1960) was unable to detect any penetration by phosphotungstic acid in cellulose microfibrils, which suggested that these elementary fibrils were crystalline along their entire length. The microfibrils examined in this study appeared to be composed of elementary fi-

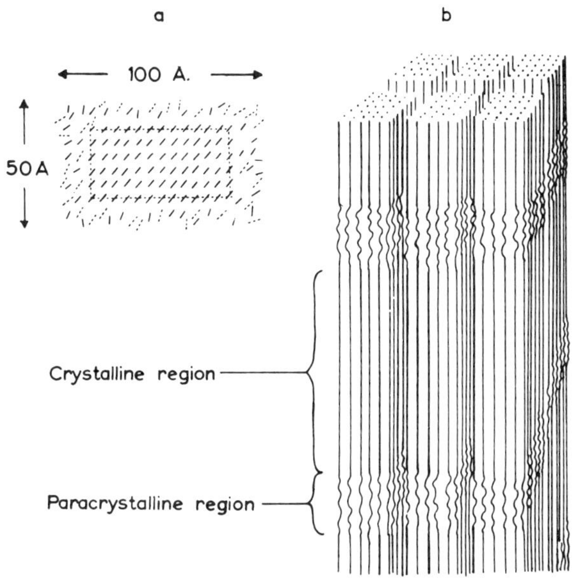

Fig. 7.4. (a) Microfibrillar structure according to Preston and Cronshaw (1958). (b) Model of cellulose fibrillar structure postulated by Hess *et al.* (1957). Reprinted from "Cellulases and Their Applications." Cowling and Brown (1969).

brils, each having a width of about 3.5 nm and containing 36 cellulose molecules. Mühlethaler (1960) proposed the model shown in Fig. 7.5, which incorporated the antiparallel nature of the cellulose chains predicted in the Meyer–Misch structure (Meyer and Misch, 1937). In this model 16 of the 18 pairs of antiparallel chains were at the surface, and the only disorder or disturbance in this structure was the chain-end dislocation. The X-ray scattering caused by this disturbance in synthetic fibers was similar to that previously attributed to the amorphous regions (Mühlethaler, 1969).

Further studies on the nature of the elementary fibrils were carried out by Manley (1964), who referred to them as protofibrils. He prepared aqueous suspensions of microfibrils from ramie and other fibers by ultrasonic disintegration and examined them under the electron microscope using a negative staining technique. The microfibrils were composed of protofibrils 3.5 nm in diameter and resembled a string of pearls due to a periodic variation in structure. According to Manley (1964), this effect was caused by the cellulose molecules in the protofibrils being in a folded-chain structure and forming a ribbon 3.5 nm wide. When tightly wound,

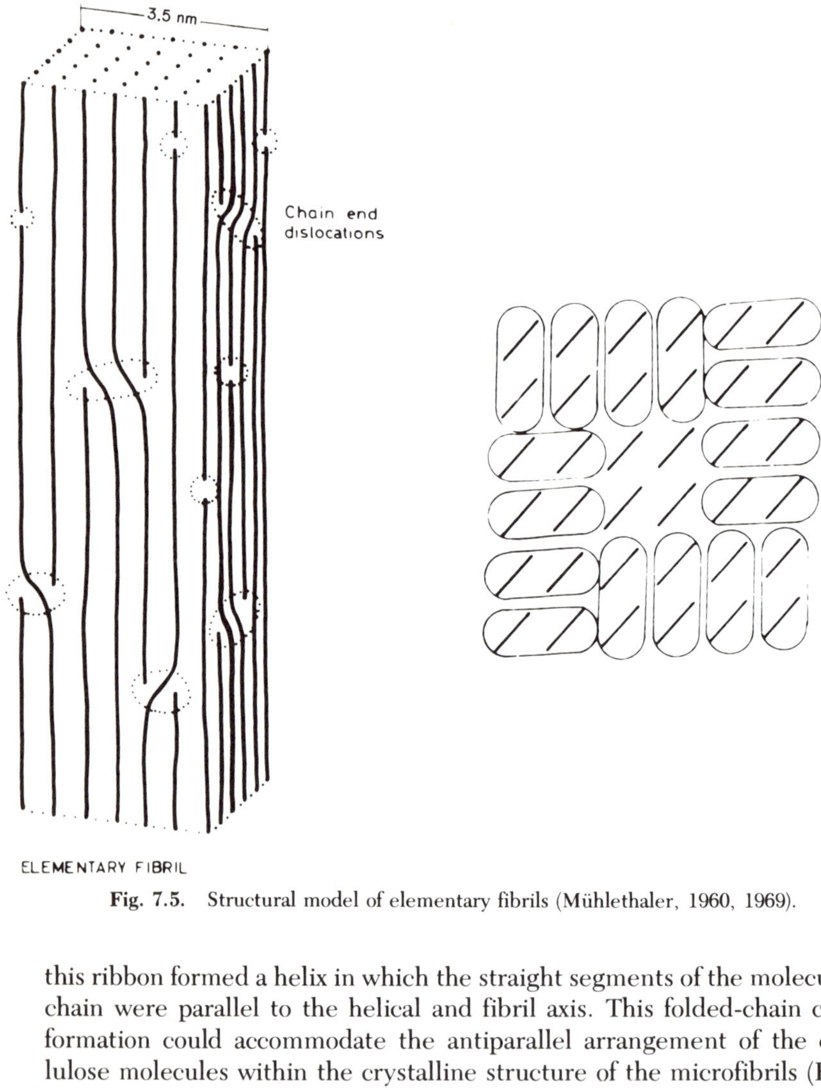

Fig. 7.5. Structural model of elementary fibrils (Mühlethaler, 1960, 1969).

this ribbon formed a helix in which the straight segments of the molecular chain were parallel to the helical and fibril axis. This folded-chain conformation could accommodate the antiparallel arrangement of the cellulose molecules within the crystalline structure of the microfibrils (Fig. 7.6a). Marx-Figini and Schultz (1966) proposed a modification of Manley's model in which the folded length was 15–20 nm instead of 3.5 nm (Fig. 7.6b).

The existence of the elementary fibril was challenged by several researchers who considered this subdivision of the microfibril to be an artifact arising from sample preparation for electron microscopy (Nieduszynski and Preston, 1970; Preston, 1971). The various techniques used to

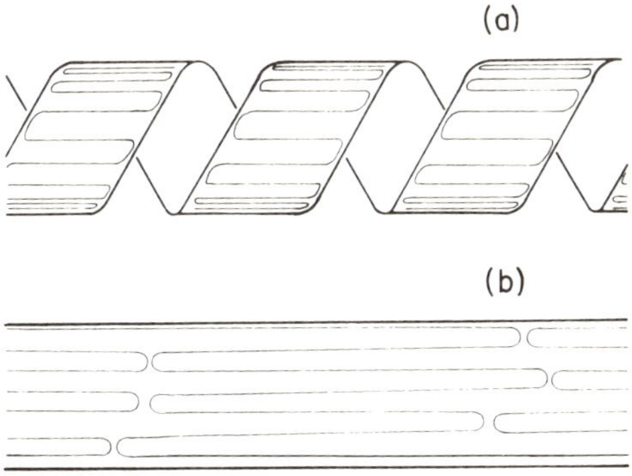

Fig. 7.6. Folded-chain structures of protofibrils (a) according to Manley (1964) and (b) according to Marx-Figini and Schultz (1966).

establish the presence of elementary fibrils, including negative staining, were considered to be extremely unreliable and misleading. These researchers examined the crystal dimensions of a variety of microfibrils using low-angle X-ray scattering and high-angle line-broadening techniques together with electron microscopy. They were unable to confirm the existence of the elementary fibril due to significant differences between their intensity data in ramie fibers and the data reported by Mann *et al.* (1960). Preston (1971) therefore questioned the existence of the elementary fibrils and suggested that the basic biological unit of cellulose was the microfibril. This varied in transverse dimension from 10 × 15 nm to 30 × 15 nm and contained a single central crystalline core.

A further controversy developed regarding the orientation of the constituent cellulose molecules within the microfibrils. This appeared to be resolved by a study of Muggli *et al.* (1969) undertaken to determine whether the cellulose molecules were in an extended form or folded-chain arrangement. These researchers transversely sectioned ramie fibers into 2-μm sections with a microtome knife and examined the molecular weight distribution curve of these sections by gel permeation chromatography. As can be seen in Fig. 7.7 if the arrangement were an extended one, many of the molecular chains would be cut randomly, resulting in a marked reduction in molecular weight. In the case of a folded-chain structure, however, relatively few microfibrils would be cut so that the majority would retain their original molecular weight. These researchers observed a decrease in molecular weight from 3900 to 1600, which indicated that the cellulose molecules were in the extended form. This was consistent with the fringe-micellar theory, which formed the basis of the Mühlethaler model. Holt and Sellen (1975) questioned the ability of this technique to

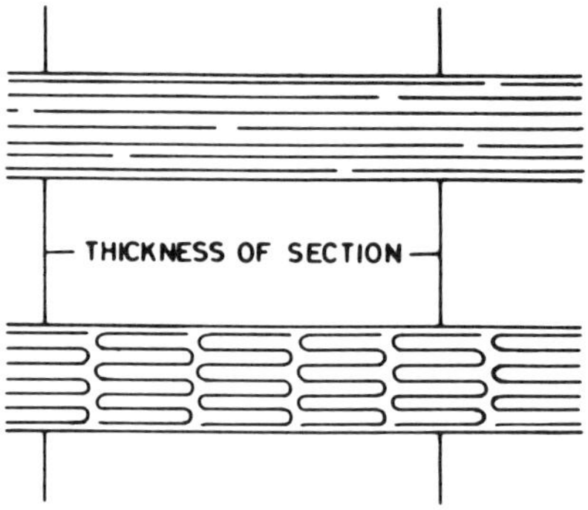

Fig. 7.7. Sectioning of cellulose fibers having an extended (top) or folded-chain (bottom) conformation (Muggli *et al.*, 1969).

disprove the folded-chain hypothesis. They determined the statistics of chain sectioning on the chain length distribution of cellulose. Many difficulties were encountered in obtaining transverse sections of uniform thickness as well as in subsequently preparing fully soluble and substituted derivatives for gel chromatography. These workers were unable to find any simple relationship between gel permeation chromatography of the high molecular weight cellulose derivatives and the molecular weight distribution curves. These results indicated that the folded-chain arrangement of cellulose could not be ruled out by means of this technique, as suggested by Muggli *et al.* (1969). It is apparent from these studies that further research is required to establish the arrangement of cellulose molecules.

III. CELLULASES

The degradation of cellulose is catalyzed by cellulases (β-1,4-glucan glucanohydrolase, EC 3.2.1.4). These enzymes play an important physiological role in the abscission of fruits and leaves (Abeles, 1969; Rasmussen, 1973, 1974, 1975; Rasmussen and Jones, 1971; Young, 1972). Their contribution to the softening of fruits, however, remains to be clarified. Hall (1963, 1964) monitored the activity of cellulase during the ripening of tomatoes. An increase in activity was reported during the development of the tomato fruit, with the locular material being markedly higher in cellulase than the pericarp. Further research indicated that this enzyme may complement pectic enzymes in fruit softening (Dickinson and McCollum,

1964; Maclachan and Perrault, 1964; Halla nd Mullins, 1965). An increase in cellulase activity during the ripening of tomatoes was also observed by Hobson (1968), who was unable to correlate this with a loss of firmness. This was based on the fact that the tomato species from the subgenera *Eriopersicon* and *Eulycopersicon* were similar in firmness in spite of a markedly higher cellulase activity in the subgenus *Eulycopersicon*. These results were consistent with studies on pears (Jeremyn and Isherwood, 1956) and peaches (Sterling, 1961) which suggested that cellulase did not contribute to fruit ripening.

The relationship between firmness and cellulase activity of several tomato lines was examined by Sobotka and Watada (1971) at different stages of maturity. An increase in enzyme activity was reported for all the tomato lines as the fruit softened during ripening, although the level of cellulase activity varied considerably (Fig. 7.8). In all cases cellulase activity reached a maximum 4 days after the breaker stage during storage at 22°C. The commercial varieties Manhattan and West Virginia (W. Va. 63) exhibited enzyme activity two to three times higher than that of the experimental T lines (T3640, T3585, and T3624). When cellulase activity was compared to firmness as measured with the Pressure Load Firmness Meter (Foda, 1957), the tomato lines with the lower cellulase activity were firmer. Thus, T3640 with the lowest cellulase activity was the firmest whereas West Virginia 63 with the highest cellulase activity developed softness far more rapidly. Significant correlation coefficients of 0.69–0.79 were reported by these researchers between enzyme activity and firmness, which supported a role for cellulase in fruit softening at a specific stage of maturity. The discrepancy between these results and those of Hobson (1968) was attributed to differences in the tomato cultivars used. Sobotka and Watada (1971) commented, however, that cellulase was probably not the sole enzyme responsible for the softening of the tomato fruit. This was based on the observation that the sharp decline in firmness during the initial stages of ripening was accompanied by insignificant increases in cellulase activity. These initial changes in firmness were attributed to the pectolytic enzymes.

Similar studies on cellulase activity were reported by Hasegawa and Smolensky (1971) during the ripening of Deglet Noor dates. They reported a dramatic rise in enzymatic activity, which leveled after the late red stage, as shown in Fig. 7.9. This increase in cellulase activity paralleled the increase in polygalacturonase activity observed previously by Hasegawa *et al.* (1969) during the initial stages of ripening characteristic of climacteric fruits. A close relationship was evident between the levels

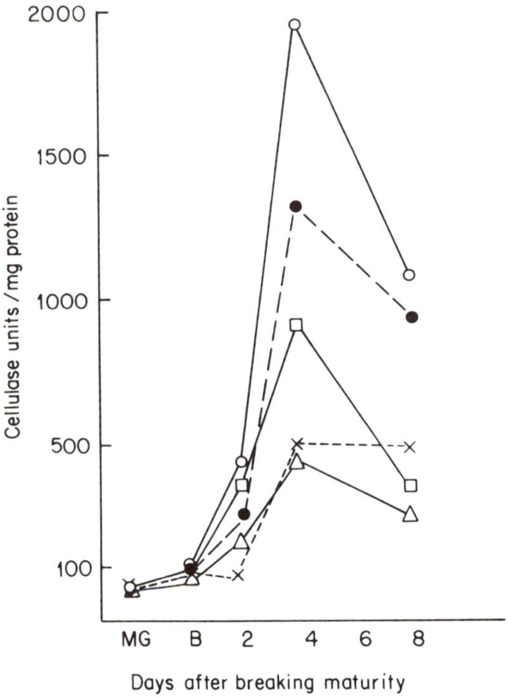

Fig. 7.8. Cellulase activity of several ripening tomatoes. Key: ○, W. Va. 63; ●, Manhattan; X, T3640(hpcc); □, T3585(cc); △, T3624(hp); MG, mature green; B, breaking. (Sobotka and Watanda, 1971). Breeding lines designated as the "T-lines" that contain genes for high pigment (hp), crimson characteristic (cc), or both (hpcc).

of these enzymes and the softening of the dates. Hasegawa and Smolensky (1971) showed that the mature date had considerably more cellulase activity than the immature date. The increase in degradation products of the cell wall polymers of Deglet Noor dates during the late red stage was attributed to the hydrolytic enzymes, including cellulase (Coggins and Knapp, 1969).

A. Nature of Cellulase

Considerable research has been carried out on microbial cellulases, whereas plant cellulases have been poorly researched until fairly recently. The degradation of cellulose by microbial cellulolytic enzymes involves the combined action of several distinct enzymes referred to as the "cel-

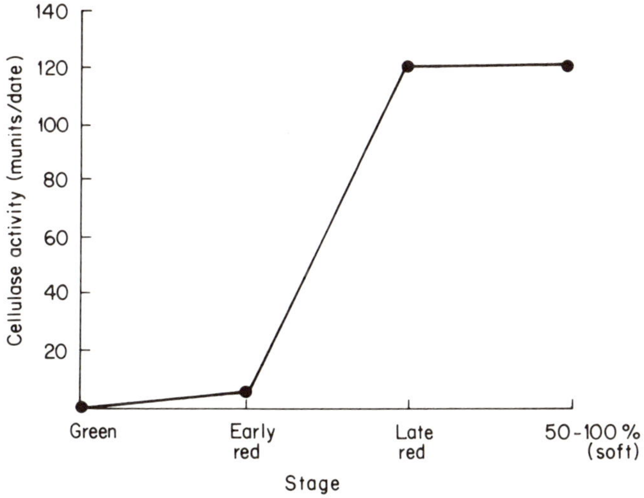

Fig. 7.9. Cellulase activity in Deglet Noor dates at different stages of maturity. (Adapted from Hasegawa and Smolensky, 1971.)

lulase complex" (King and Vessal, 1969; Li *et al.*, 1963; Reese *et al.*, 1950). These include C_1-cellulase, C_X-cellulase, cellobiase, and exocellulase. Of these C_1-cellulase is the least understood, although it is thought to initiate the degradation of insoluble cellulose to soluble derivatives. C_X-Cellulase, which then degrades the soluble cellulose, has been assigned the systematic name 1,4-glucan 4-glucanohydrolase. This enzyme is an endoenzyme that randomly cleaves internal linkages in the cellulose chain. Cellobiase, which then hydrolyzes dimers of glucose, is the major glucosidase (EC 3.2.1.21) involved in cellulose degradation. These enzymes appear to catalyze the following sequence of reactions:

$$\begin{array}{c}\text{Insoluble cellulose}\\ \downarrow \text{ c}_1\text{-cellulase}\\ \text{Soluble cellulose derivatives}\\ \downarrow \text{ c}_x\text{-cellulase}\\ \text{Cellobiose}\\ \downarrow \text{ cellobiase } (\beta\text{-1,4-glucosidase})\\ \text{Glucose}\end{array}$$

Pharr and Dickinson (1973) carried out an extensive study to characterize cellulase from the locular tissue of ripening tomato fruit. Two enzymes were identified, C_X-cellulase and cellobiase, both of which exhibited properties that were similar to those of cellulolytic enzymes of fungal origin (King and Vessal, 1969). The combined effect of these enzymes could explain the ability of the tomato preparation to partially degrade carboxymethylcellulose to glucose. However, the absence of C_1-cellulase or exocellulase in their preparations raised doubts about the ability of the tomato extract to degrade insoluble cellulose. This was resolved by Sobotka and Stelzig (1974), who partially purified four enzyme fractions from 2-day, postbreaker tomatoes (*Lycopersicon esculentum* L., commercial variety Manhattan) by ammonium sulfate fractionation. Two of these fractions (I and IV) appeared to be characteristic of C_X-cellulase, whereas the other two fractions (II and III) were more typical of β-glucosidases (Fig. 7.10). In addition to hydrolyzing carboxymethylcellulose the tomato enzymes (fractions I and IV) were also capable of degrading insoluble cellulose materials. This indicated a C_1-cellulase type of enzyme in the tomato fruit, although these researchers suggested that this reaction could be catalyzed by one or more of the C_X-cellulase components. Fraction IV appeared to be more typical of C_1-cellulase than fraction I because it could release large products from insoluble cellulose material but was incapable of degrading short-chain cellodextrins. Fraction II was tentatively classified as a nonspecific β-glucosidase since it was active on a number of different glucoside substrates. Fraction III differed from fraction II in being extremely active toward cellotetraose and inactive toward *p*-nitrophenyl-D-glucoside. The main enzyme in fraction III was considered to be exo-1, 4-glucanase. This work is important because it demonstrates the ability of tomato cellulase to completely degrade insoluble cellulose to glucose.

B. Growth Regulators and Cellulase Activity

The ability of growth regulators to stimulate the cell-wall-degrading enzymes in plants is well established (Datko and Maclachan, 1968; Tanimoto and Masuda, 1968). The effects of several growth regulators on the activities of cellulase and polygalacturonase were investigated by Babbitt and co-workers (1973) in ripening tomatoes. Extreme care was taken in selecting tomato fruit of known age since these workers considered this to be a better indication of maturity than the criteria of size and color used by Hobson (1965. 1968). Cellulase activity was detected in the very young fruit but decreased midway through the development of the fruit and then increased steadily during maturation. Polygalacturonase activity was not observed until the color of the tomato started to change. The effect of the growth regulators ethephon and gibberellic acid on cellulase and polygalacturonase activity is shown in Fig. 7.11. In the presence of ethephon, cellulase activity increased initially but then de-

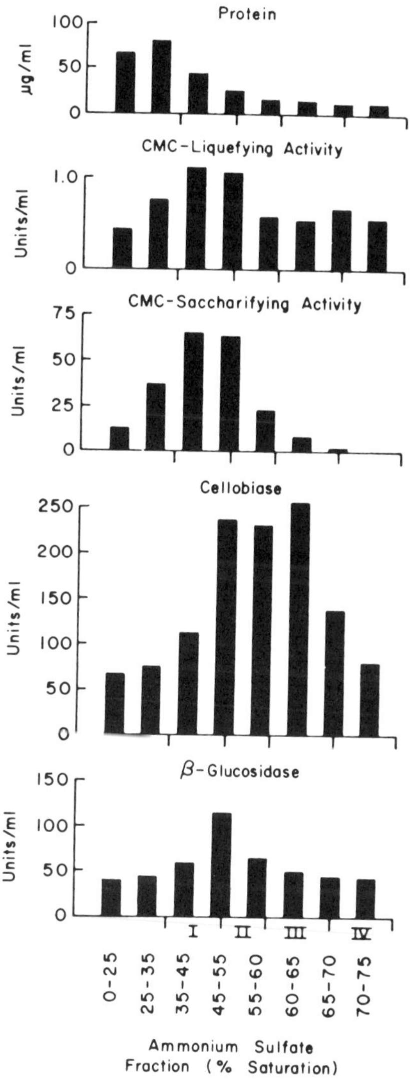

Fig. 7.10. Ammonium sulfate fractionation of total protein, CMC-liquifying, CMC-saccharifying, cellobiase, and aryl-β-glucosidase activities of a crude extract of 2-day post-breaker Manhattan commercial variety tomato fruit (Sobotka and Stelzig, 1974).

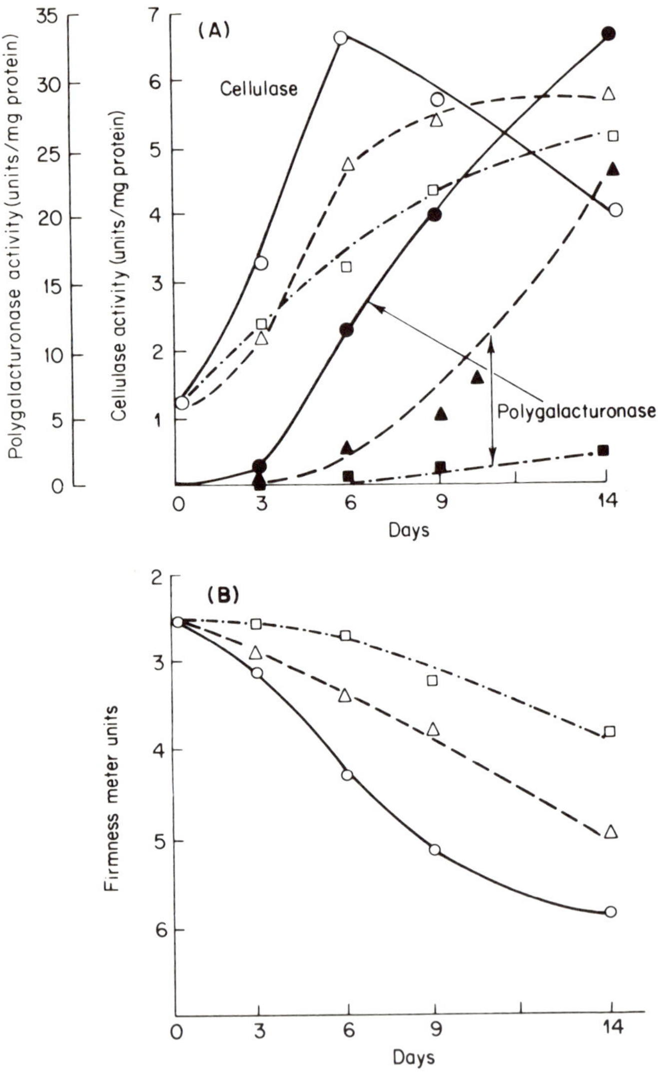

Fig. 7.11. Effect of ethephon (○) and gibberellic acid (□) on enzyme activity (A) and firmness (B) of tomatoes (control, △) (Babbitt *et al.*, 1973).

clined after 6 days, whereas polygalacturonase activity increased. Gibberellic acid suppressed polygalacturonase activity almost totally, whereas cellulase activity increased steadily. This was accompanied by a decrease in the overall firmness of the fruit, which provided convincing evidence for the role of cellulase in fruit softening. Babbitt and co-workers (1973) suggested that cellulase initiated fruit softening by degrading the cellulose fibrils of the cell wall to permit the pectic enzymes to penetrate the middle lamella.

IV. HEMICELLULOSES

Hemicelluloses are a rather poorly defined group of noncellulosic, nonpectic cell wall polysaccharides readily extracted with alkaline solution. They are important constituents of the primary and secondary walls of plants and appear to link the pectic and cellulosic chains (Bauer *et al.*, 1973; Northcote, 1969). Together with cellulose, lignin, and pectic substances, hemicellulose forms the major portion of undigested carbohydrates in food, which are referred to as dietary fiber. Hemicellulose is composed of three main groups of polysaccharides: the xylans, the mannans and glucomannans, and the galactans and arabogalactans.

A. Xylans

Xylans are widely distributed in higher plants and consist of polymers of β-1,4-linked xylopyranose chains. Xylans are characterized by their side chains, such as β-3-linked arabinofuranose in corn cobs and wheat straw. Chanda *et al.* (1951) reported that glucuronic acid in pears is linked α-1,3 to the xylan chain.

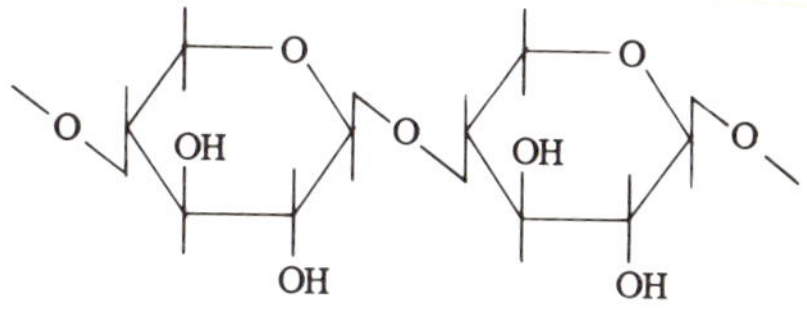

Xylan (repeating unit)

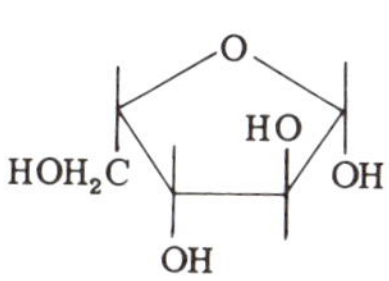

L-Arabinofuranose

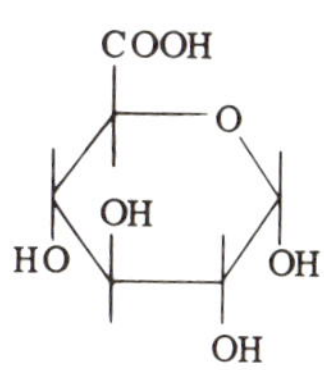

D-Glucuronic acid

Bauer *et al.* (1973) isolated xyloglucan, a β-1,4-linked glucan with terminal branches of α-1,3-linked xylopyranose, from the walls of suspension-cultured sycamore cells. These researchers found hemicellulose to be covalently linked to pectic material and noncovalently to cellulose fibrils. They suggested that hemicelluloses should be redefined as those plant polysaccharides that are noncovalently bound to cellulose.

B. Mannans and Glucomannans

These polymers are derivatives of β-1,4-linked mannopyranose and β-1,4-linked glucopyranose units with various proportions of glucose and mannose. The ratio of glucose and mannose depended on the source, although for coniferous and deciduous woods these were 1 : 3 and 1 : 2, respectively (Aspinall, 1970).

C. Galactans and Arabogalactans

Galactans and arabogalactans are highly branched derivatives of 1,3- and 1,6-linked D-galactopyranose polymers with arabinose side chains. The majority of arabinose is present in the form of a disaccharide (3-*O*-L-arabopyranosylarabofuranose) linked via C-6 of the galactopyranose chain. Galactan, although not completely free of arabinose, is still referred to by this name since the major proportion of residues are galactose. However, whenever larger amounts of L-arabinose are present, this polymer is referred to as arabogalactan.

V. HEMICELLULASES

The enzymatic degradation of hemicelluloses is poorly understood. Research in this area has been concerned primarily with the infection of plants by phytopathogenic organisms capable of degrading the complex polysaccharides of the plant cell wall (Decker and Richards, 1976).

Chapter 8

Pectic Substances

I. INTRODUCTION

Pectic substances consist of a mixture of three polysaccharides: galactans, arabans, and rhamnogalacturonans (Aspinall, 1970). Galactans are long, linear polymers of β-1,4-linked D-galactopyranose units, whereas araban is a low molecular weight polysaccharide comprised of L-arabofuranose residues, some of which are α-1,5 linked and some of which are α-1,3 linked.

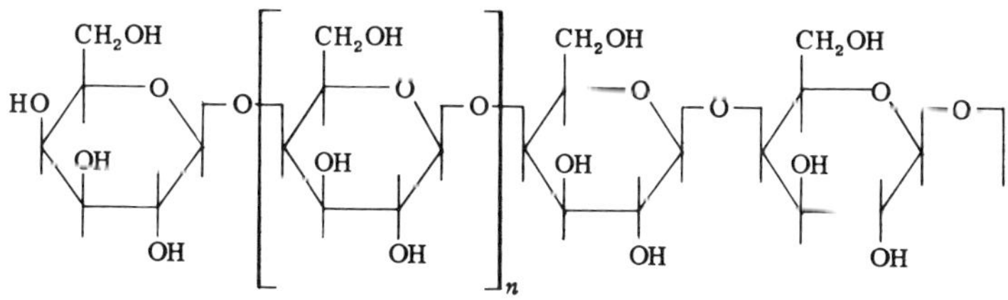

Galactan (n = 120)

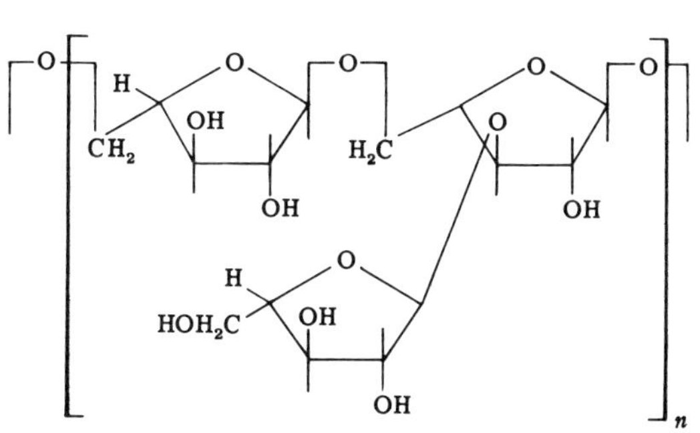

Araban

The predominant polysaccharide of pectic substances is rhamnogalacturonan, a high molecular weight polymer composed of a backbone of α-1,4-linked D-galacturonopyranose residues interspersed with α-1,2-linked rhamnopyranose. As many as 65% of the carboxyl groups of poly-D-galacturonic acid are esterified by methyl groups, whereas positions 2 and 3 might be acylated. Because of the predominance of galacturonan the properties of pectic substances are determined by this polysachararide.

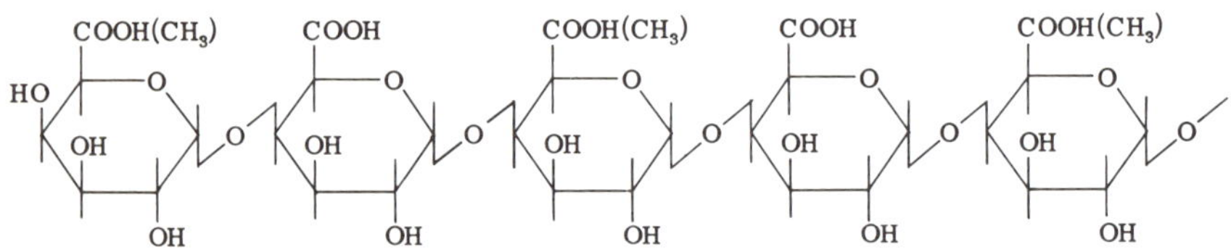

α-1,4-Linked polygalacturonic acid

II. CELL WALL COMPONENTS

Pectic substances are the major constituents of the middle lamella as well as structural elements in the primary cell wall (McClendon, 1964; Talmadge *et al.*, 1973). They comprise a substantial portion of the cell walls of soft tissues such as the parenchymatous tissues of soft fruits and fleshy roots. At one time the three major pectic components were considered to be held together within the cell wall by forces of mechanical mixing. Talmadge *et al.* (1973) suggested that the pectic substances are covalently linked by the neutral sugars galactan and araban, which serve as bridges between rhamnogalacturonan and the hemicellulose wall components. The majority of these covalent linkages were attributed to esterification between some of the carboxyl groups of the galacturonan polymer and the hydroxyl groups of araban and galactan.

The pectic substances in the cell wall were reported by Reeve (1970) to fill the spaces between the microfibrils as amorphous, encrusting substances. Several researchers, however, provided indirect evidence for the existence of pectin elementary fibrils in the form of stable aggregates of pectic molecules (Palmer *et al.*, 1947; Reese, 1969; Roelofsen and Kreger, 1951). Using positive and negative staining techniques, Leeper (1973) confirmed the existence of pectate elementary fibrils as suggested by Roelofsen and Kreger (1951) for collenchyma cell wall pectin. These fibrils were 30.8±6.2 Å in width and contained 13 pectate molecules in their cross-sectional areas. A decrease in intermolecular distance accounted for the smaller width observed in the dehydrated material.

The pectic substance of the primary wall is protopectin, a generic name given to the water-insoluble parent compound of the pectic substances in unripe fruit. In spite of an excellent review on protopectin by Joslyn (1952), relatively little is known about this compound. It was suggested that there is a greater degree of esterification of the galacturonan chain in protopectin, minimizing the degree of salt formation, than in pectic substances in the middle lamella. The hydrolysis of protopectin in plants leads to the formation of pectin or pectinic acids (Eskin *et al.*, 1971; Worth, 1967). These are polygalacturonic acid polymers possessing varying amounts of methyl esters, which can be removed by hydrolysis or by enzymes to form pectic acid free of methyl esters.

The pectic substances in the middle lamella are distinct from those in the primary and secondary cell walls. This is apparent from the different responses to staining and from the failure of hydrochloric acid to extract pectic substances from the middle lamella. The middle lamella, an amorphous region between the primary cell wall and adjacent cells, is composed mainly of pectic substances either in the free state or as calcium pectate.

III. PECTIC ENZYMES

The softening of fruit during ripening and storage is attributed to changes in pectin (Kertesz, 1951; Pressey *et al.*, 1971; Hobson, 1967; Tavakoli and Wiley, 1968). These changes are catalyzed by two groups of enzymes, pectin methylesterases (EC 3.1.1.11) and polygalacturonases (EC 3.2.1.5). Pectin methylesterases (PME), still referred to by their trivial name pectinesterases, catalyze the deesterification of pectin, yielding a uronic acid carboxyl and methanol. Polygalacturonases catalyze the splitting of 1,4 glycosidic bonds of the D-galacturonan chain in the pectin molecule (Scheme 8.1).

A. Pectinesterase

Kertesz (1938) reported a twofold increase in pectinesterase activity during the ripening of the tomato fruit. Pithawala *et al.* (1948) observed about a 33% increase in enzyme activity as the tomato matured from the green to ripe stage. A similar relationship was found by Hobson (1963)

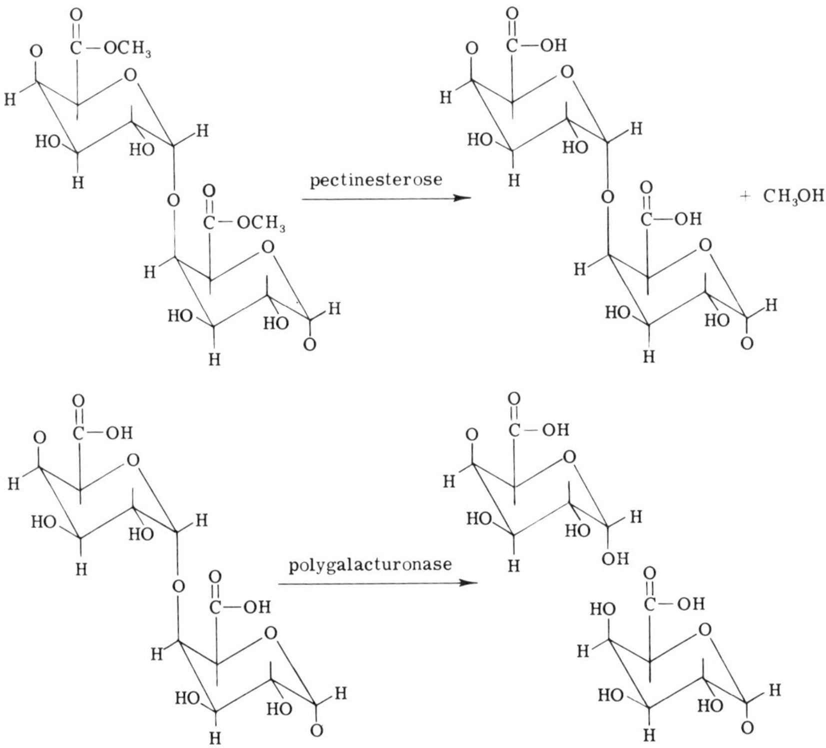

Scheme 8.1. Reaction mechanisms of pectinesterase and polygalacturonase.

when he monitored a 40% increase in enzyme activity as the tomato fruit progressed from mature green to red ripe. Hamson (1952) noted that the firmer varieties of tomatoes exhibited much higher pectinesterase activity. On the basis of these results, Hobson (1963) predicted that the tomato variety Immuna would have a lower pectinesterase activity than Potentate since it was less firm at all stages of ripening. However, the opposite results were obtained. Previous studies by Hobson (1962) and Hall and Dennison (1960) also failed to establish any correlation between the activity of this enzyme and firmness of tomato fruits.

Lee and Macmillan (1968) and Delincee and Radola (1970) suggested that the activity of pectinesterase in tomatoes was due to a single enzyme. Later work by Pressey and Avants (1972) confirmed the presence of multiple forms of pectinesterase in tomatoes. These fractions were not identified previously because of their low activity, the similarity in their molecular weights, and the difficulty in separating them by gel filtration. The major fraction isolated by these researchers from their tomato extracts as well as from a commercial sample of pectinesterase was referred to as PE IV. This fraction was similar to the enzyme discussed by Lee and Macmillan (1968) and was considered to be the most important pectinesterase enzyme fraction in tomato extracts. Differences were evident in the number of enzyme fractions and activities among varieties and at different stages of maturity. The presence of more than one pectinesterase in tomatoes probably reflects the complexity of pectin and the many changes it undergoes during fruit development. One of the pectinesterases in tomatoes was considered to create initiation points for other enzyme fractions.

An increase in pectinesterase activity was reported in Doyenne Boussoch pears by Weurman (1954). This reached a maximum during the initial stages of development and then decreased rapidly. Weurman monitored pectinesterase activity by measuring the increase in viscosity of a pectic solution. However, since polygalacturonase activity was also present these results are questionable. Nagel and Patterson (1967) monitored the activity of this enzyme more carefully during the development of pears (*Pyrus communis*). When pectinesterase activity was expressed as specific activity (PME units/mg dry protein), it decreased during maturation of the fruit. Interference by tannins or phenolic compounds was eliminated by extraction of the enzyme in the presence of polyvinylpyrrolidone (PVP). When this treatment was applied, the activity of the pectinesterase preparations all increased, although the overall trend was a decrease from ripening to the overmature stage (Fig. 8.1).

In contrast to pectinesterase activity reported in pears, Hultin and Levine (1965) observed an increase in the activity of this enzyme in the pulp of ripening bananas. Later studies by de Swardt and Maxie (1967), however, showed by the removal of polyphenols with PVP that the activity of this enzyme remained constant throughout the ripening period. Palmer (1971), using similar methods to remove polyphenols, also reported no change in pectinesterase activity in the pulp of ripening bananas. Further confirmation was provided by Brady (1976), who found little change in pectinesterase activity in the presence of 2-mercaptobenzothiazole, a potent inhibitor of banana polyphenol oxidase activity (Palmer and Roberts, 1967). These results contradicted the work of Hultin and Levine (1965), who did not remove the effects of polyphenols. Further doubts were expressed by Brady (1976) regarding the behavior of pectinesterase during ion-exchange chromatography as reported by Hul-

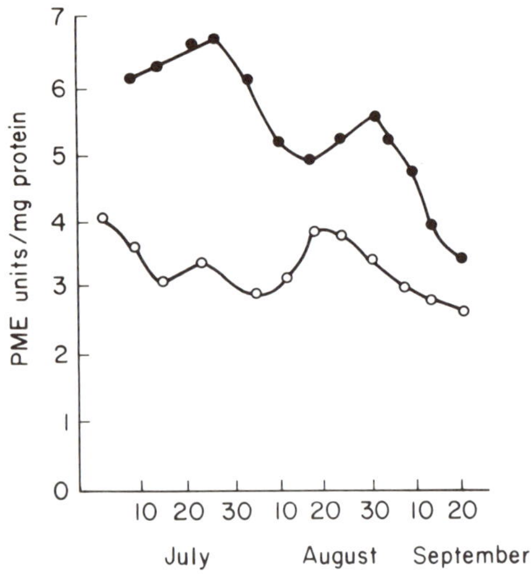

Fig. 8.1. Specific activity of pectinesterase in the pear during maturation on the tree (○, 1962, $r = -0.54$; ●, 1963, $r = -0.84$) (Nagel and Patterson, 1967).

tin and Levine (1965), in which the changes observed probably did not reflect the changes *in vivo*.

Zaubermann and Schiffman-Nadel (1972) monitored pectinesterase activity during the ripening of Fuerte avocados. Although they did not mention any precautions taken to remove polyphenols from their preparations, they nevertheless noted a decrease in the activity of the enzyme. As can be seen in Fig. 8.2, this decrease was pronounced in the young fruit (June/July) but was gradual in the mature fruit (September/March). Fruit softening was observed in the young and mature fruit when pectinesterase activity was minimal. This precluded a direct role for this enzyme in the ripening process. The decline in pectinesterase activity before the softening of the avocado was also observed by Rouse and Barmore (1974).

B. Pectinesterase Activity as an Index of Avocado Maturity

Zaubermann and Schiffman-Nadel (1972) found that the avocado fruit reached a commercial picking size when the pectinesterase activity was below 105 PME units. They suggested that this drop in pectinesterase activity might be used to estimate the maturity of the fruit before the picking season. Barmore and Rouse (1976) suggested that it might be used

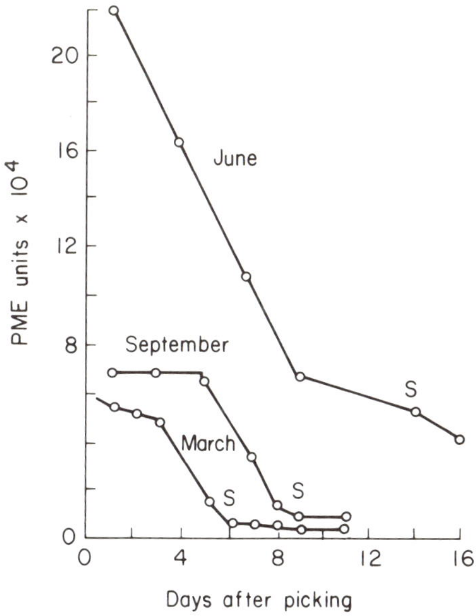

Fig. 8.2. Pectinesterase activity during storage of avocados at various stages of development (S denotes softening; fruits stored at 20°C) (Zaubermann and Schiffmann-Nadel, 1972).

to assess the softening time of avocados. Their study was conducted on avocados subjected to controlled-atmosphere storage. This method was shown by Spalding and Reeder (1972) to increase the storage life of avocados from 20 to 60 days without decreasing the softening time of the fruit. Barmore and Rouse (1976) showed that pectinesterase activity is a reliable indicator of the softening time of avocados stored in a controlled atmosphere. In some of the stored avocados the level of pectinesterase was close to that observed in normal soft fruit, yet the avocados remained firm for at least a day. This was attributed to inhibition of polygalacturonase activity by the controlled-atmosphere storage conditions. Therefore, this provides further evidence for the indirect role of pectinesterase in the softening of fruits.

C. Polygalacturonase

Polygalacturonases (PGA) are more clearly implicated than pectinesterases in the process of fruit softening (Hasegawa *et al.*, 1969; Hobson, 1965; Pressey *et al.*, 1971; Zaubermann and Schiffman-Nadel, 1972).

Hobson (1964) did not detect PGA activity in green tomato fruit but found the activity to increase exponentially with ripening. She found PGA close to the surface of the ripe fruit, as observed by McColloch *et al.* (1952). A close relationship between the activity of this enzyme and ripening was proposed by Hobson (1964) when the tomato variety Immuna, which was less firm than the variety Potentate, was found to be higher in pectolytic enzyme activity. This was consistent with the previous findings of Foda (1957), who reported that of the two varieties studied, Stoledale and Y13, the softer one possessed greater PGA activity. This relationship received further support from Hobson (1965), who found that an increase in firmness in six varieties of tomatoes was accompanied by a corresponding decrease in PGA activity.

D. Nature of Polygalacturonase

Luh *et al.* (1956) studied the properties of a partially purified PGA fraction from tomatoes. When added to pectic acid at pH 4.5, it caused a 4% hydrolysis of glycosidic bonds, which was accompanied by a 50% decrease in viscosity. The reaction proceeded rapidly at a linear rate to around 30% hydrolysis with only oligomers detected as products. Further hydrolysis occurred at a much slower rate, but it was not until 80% hydrolysis had occurred that tri-, di-, and monogalacturonic acids were detected. Patel and Phaff (1960a,b) carried out a more extensive purification of tomato PGA, obtaining a sixfold increase in the specific activity. Two pH optima of 4.5 and 2.5 observed when the substrate was soluble pectic acid suggested the presence of more than one enzyme, as previously indicated by Foda (1957) who reported pH optima of 3.5 and 5.0 for his PGA extract in the presence of pectic acid. Increased reaction velocity at the lower pH was attributed to a more rapid hydrolysis of the oligomers, as observed by Demain and Phaff (1957) for yeast PGA. The various pH optima reported are summarized in Table 8.1. These data point to the presence of several PGA isoenzymes in tomatoes of different substrate size specifications.

Pressey and Avants (1971) examined the relationship between substrate size and tomato activity. They found that the enzyme reached a pH optimum of 5.0 in the presence of the high molecular weight polymer pectic acid. Below pH 4.0 the activity was negligible. As the molecular weight of the substrate decreased the pH optimum shifted to the acid side. Addition of NaCl enhanced PGA hydrolysis of the various substrates studied. This effect, more pronounced at low pH, was dependent on the

TABLE 8.1

pH Optima of Tomato PGA

pH optimum	Substrate	Reference
2.5	Short-chain pectic acid	Patel and Phaff (1960a,b)
3.5	Normal pectic acid	Foda (1957)
4.5	Normal pectic acid	Luh *et al.* (1956), Patel and Phaff (1960a,b), Roelofsen (1953)
5.0		Foda (1957), Pressey and Avants (1971)

molecular size of the substrate. The hydrolysis of the lower molecular weight substrates by PGA was inhibited 70% in the presence of pectic acid. This suggested the presence of at least two PGA isoenzymes, one of which was not inhibited by high molecular weight polygalacturonic acid.

Pressey and Avants (1973) separated two PGA fractions from ripe tomatoes by chromatography on DEAE-Sephadex A-50, as shown in Fig. 8.3. The major fraction, PGA II, was similar to PGA I but was far more

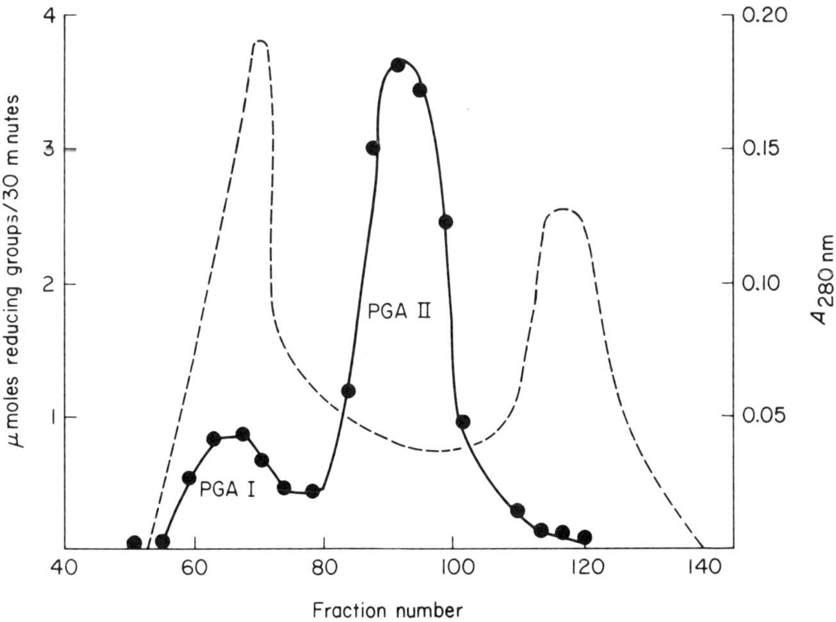

Fig. 8.3. Chromatograph of Tomato Polygalacturonase on DEAE-Sephadex A-50 (●——●). Polygalacturonase activity; ----, protein concentration) (Pressy and Avants, 1973).

heat labile. The PGA II fraction was completely inactivated at 65°C compared to 90°C for PGA I. The PGA II fraction also exhibited maximal stability at pH 5.6 compared to pH 4.5 for PGA I. The activity of both fractions shifted to the acid side as the molecular weight of the substrate decreased. Only PGA I hydrolyzed pectic acid in the absence of NaCl, although maximal activity was reached in the presence of 0.2 *M* NaCl. This contrasted with PGA II, which required the presence of 0.3 *M* NaCl before enzyme activity was detected. By means of gel filtration the molecular weights of PGA I and PGA II were found to be 84,000 and 44,000 respectively. The rapid reduction of the viscosity of pectic acid by PGA II compared to PGA I indicated to Hunter and Elkan (1974) that PGA I was probably an exopolygalacturonase and PGA II an endopolygalacturonase. When these researchers extracted PGA from ripening tomato fruit, they obtained a 44-fold purification following a separation of the dialyzed 50–80% ammonium sulfate fraction on a Sephadex G-100 column. The enzyme was active toward pectic acid but not toward pectin. A low level of pectic acid hydrolysis by this enzyme is characteristic of an endo or random attack since a 2.2% hydrolysis caused a 94% decrease in viscosity (Table 8.2). Wallner and Walker (1975) obtained a PGA fraction from ripening tomatoes and estimated its molecular weight to be 40,000. This was close to the 44,000 reported by Pressey and Avants (1973) for PGA II. These researchers observed only one peak following a separation of their extract on Sephadex G-75 or G-100. The absence of a

TABLE 8.2

Relationship between Decrease in Viscosity and Hydrolysis by Tomato Endopolygalacturonase[a]

Reaction time (sec)	Decrease in viscosity (%)	Hydrolysis[b] (%)
0	0	0
5	23	−0.1
10	40	0
15	49	0.1
30	72	0.3
60	82	0.5
120	91	1.0
300	94	2.2

[a] From Hunter and Elkan (1974).

[b] Followed by the increase in reducing groups.

second peak was attributed to differences between their extraction procedure and that of Pressey and Avants (1973).

Early work by McCready and McComb (1954) could not establish the presence of polygalacturonase in peaches. Although this enzyme was absent in the unripe fruit, Pressey *et al.* (1971) detected PGA in ripening peaches. This increase in activity was accompanied by an increase in the amount of water-soluble pectin and softening of fruit, as shown in Fig. 8.4. Although softening was evident before the presence of PGA activity, the peaches softened more rapidly after the development of this enzyme. The activity of peach PGA was much lower than that in tomatoes and appeared to be associated with the particulate fraction of the homogenates. Using an improved method to solubilize and concentrate PGA from peaches, Pressey *et al.* (1971) separated two distinct fractions by gel filtration on Sephadex G-100. Neither of these fractions was activated by monovalent cations, as observed for the PGA I and II fractions isolated by Pressey and Avants (1973) from tomatoes. Both PGA peach fractions were inhibited by high levels of NaCl (0.15 *M*). Peach PGA I fraction was activated by Ca^{2+} ions with optimal activity in the presence of 0.4 m*M* $CaCl_2$. This fraction appeared to be more typical of an expolygalacturonase, cleaving off monomer units from the nonreducing end of polygalacturonate. The PGA II fraction, on the other hand, degraded pectin in a random manner, releasing soluble products characteristic of

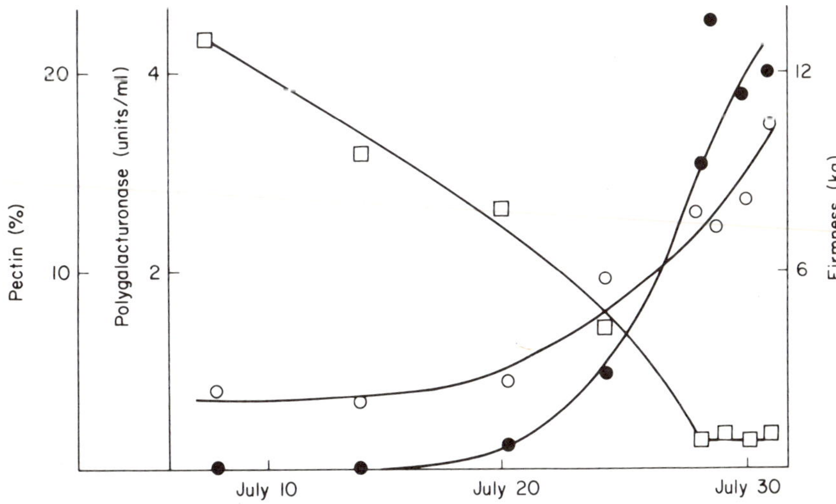

Fig. 8.4. Changes in firmness (□—□), PGA activity (●—●), and water-soluble pectin (○—○) in Elberta peaches during ripening (Pressey *et al.*, 1971).

endopolygalacturonases. The presence of both of these enzymes allows the complete solubilization of pectin to be carried out in the peach.

Pressey and Avants (1975a) reported the presence of an enzyme in carrot that was similar to the exopolygalacturonase of peach. These researchers (Pressey and Avants, 1976) also isolated endo- and exopolygalacturonases from the extracts of ripe D'Anjou pears. The main fraction was exopolygalacturonase, which was activated by the divalent cations Ca^{2+} and Sr^{2+} in a manner similar to that described for the corresponding peach enzyme. Another exopolygalacturonase was identified in cucumber by Pressey and Avants (1975b). It resembled the corresponding enzyme from peach with respect to pH optimum, cation activation, and molecular weight. Pressey and Avants (1977) isolated an exopolygalacturonase from oat seedlings that was also activated by divalent cations Ca^{2+} and Cd^{2+}.

The presence of PGA in avocados was first reported by McCready and McComb (1954). These researchers showed that a crude enzyme preparation randomly degraded pectic acid to D-galacturonic acid. Reymond and Phaff (1965) extracted PGA from the mesocarp of avocado and reported a 36-fold pruification of the enzyme. This enzyme exhibited a pH optimum of 5.5 and degraded polygalacturonic acid randomly to D-galacturonic acid, suggesting that it was an endopolygalacturonase (Demain and Phaff, 1957). Zaubermann and Schiffman-Nagel (1972) found that PGA was absent in the immature fruit while the fruit was still firm but reached a maximum as complete softening occurred (Fig. 8.5). Unlike that of pectinesterase, the activity of PGA developed in a similar manner in both the young (June/July) and more mature (September/March) fruit. The absence of detectable PGA activity in the firm fruit was attributed by Reymond and Phaff (1965) to the presence of a PGA inhibitor in the unripe avocado fruit.

E. Role of Endopolygalacturonase and Exopolygalacturonase in Fruit Softening

In tomatoes, pears, and peaches PGA activity is due to the presence of both endo- and exopolygalacturonases (Scheme 8.2). In carrots and cucumbers the exo enzyme is the sole pectate-hydrolyzing enzyme present. Endopolygalacturonase has been associated with the ripening of fruits (Hasegawa *et al.*, 1969; Hobson, 1964; Pressey *et al.*, 1971). By hydrolyzing a few internal bonds this enzyme is capable of substantially reducing pectin molecules. Since pectin plays a major role in maintaining cell wall integrity its degradation by endopolygalacturonase, following desterification by pectinesterase, softens fruit. The role of exopolygalacturonase, however, in the softening process remains unclear. This enzyme

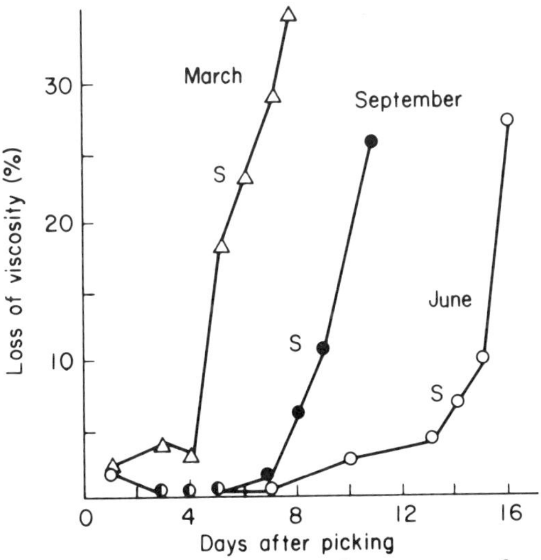

Fig. 8.5. Changes in polygalacturonase activity during storage of avocados at various stages of development (stored at 20°C; S denotes softening) (Zaubermann and Schiffmann-Nadel, 1972).

was at one time thought to continue the hydrolysis of pectate initited by endopolygalacturonase action. The affinity of exopolygalacturonase for large substrates together with its slow hydrolysis of short oligopolygalacturonates tends ot make this role doubtful. Keegstra *et al.* (1973) studied the structure of suspension-cultured sycamore cell walls. They suggested that some of the pectic components were joined to hydroxyproline-rich cell wall protein by short chains of arabinogalactan or with tetraarabinoside. Keegstra *et al.* (1973) proposed that exopolygalacturonase functions by hydrolyzing linkages between pectic polysaccharides and protein rather

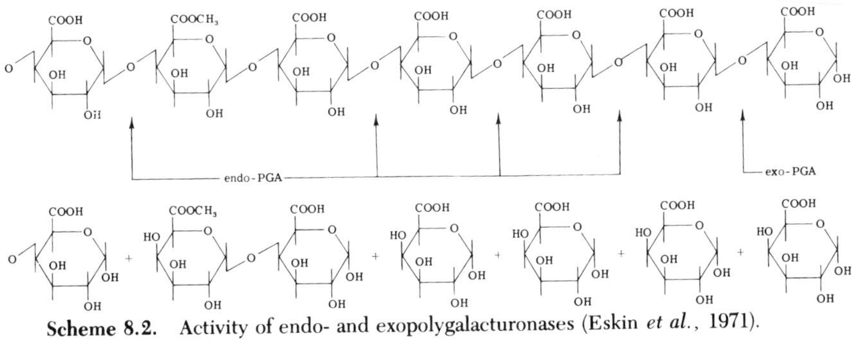

Scheme 8.2. Activity of endo- and exopolygalacturonases (Eskin *et al.*, 1971).

than by pectin degradation. Further research is necessary in order to establish the precise role of exopolygalacturonase in the fruit-softening process.

IV. PECTOLYTIC ENZYMES AND CELL WALL ISOLATES

In all the studies discussed so far, the researchers relied on model substrates to characterize the various wall-modifying enzymes. Wallner and Bloom (1977), however, felt that although such data were useful they must be interpreted with caution. For instance, does the model substrate on which the enzyme is active resemble the substrate *in situ*? If an enzyme is inactive toward model substrates does this preclude its involvement in cell wall degradation? These researchers monitored the degradation of the cell walls isolated from the green fruits of five tomato varieties, Early Red Rock, PSU 3389, Little Red, Chef, and PSU 10171, using a sodium citrate enzyme extract from ripe Early Red Rock tomatoes. The results (Fig. 8.6) indicated differences in the rate of hydrolysis of the cell wall isolates from the five tomato varieties. Of the varieties studied the isolated cell walls of Early Red Rock were the most susceptible to enzymatic degradation. This difference in susceptibility to enzymatic hydrolysis by the various cell wall isolates was attributed to differences in the amount of pectic acid or its structural arrangement in the cell wall. These results illustrated the difficulty in selecting model substrates to characterize cell-wall-degrading enzymes such as PGA whose natural substrate is so poorly defined.

Wallner and Bloom (1977) compared the degradation of cell walls during ripening with *in vitro* enzyme treatment. A 33% decrease in pericarp cell wall weight of Early Red Rock tomatoes was observed during ripening. This was close to the 28% decrease in pericarp cell wall weight observed in the *in vitro* study using Early Red Rock (green) cell walls incubated with an extract from the corresponding red ripe fruit. The similarity in wall solubilization in both of these studies demonstrated that *in vitro* wall degradation resembled the process occurring during normal fruit ripening. The *in vitro* degradation released predominantly polyuronic acid polymers together with smaller amounts of neutral sugars. The enzymes present in the citrate extract of the ripe tomato were β-galactosidase, β-1,3-glucanase and polygalacturonase. Polygalacturonase appeared to be the only enzyme capable of initiating cell wall hydrolysis. The fact that polyuronide polymers were released suggested that this enzyme was an endopolygalacturonase, as confirmed in earlier

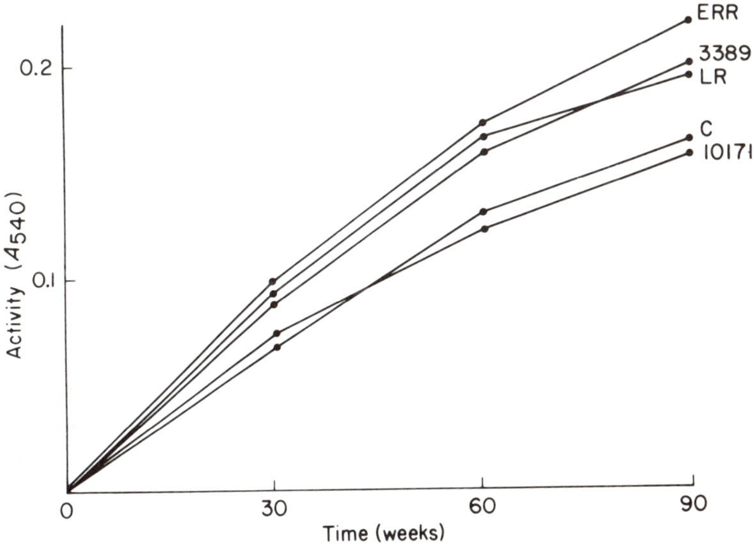

Fig. 8.6. Effect of variety on release of reducing groups from cell wall from the pericarp of green fruit. Varieties: ERR, Early Red Rock; (3389), PSU 3389; LR; Little Red; C, Chef; 1017, PSU 10171. Sodium citrate extract from red tomatoes (var. Early Red Rock) (Wallner and Bloom, 1977).

studies on tomatoes by Pressey and Avants (1973) and Hunter and Elkan (1974). The relative amounts of neutral sugars released during the *in vitro* study differed from the amounts released during normal ripening, as shown in Table 8.3. This suggested to Wallner and Bloom (1977) the presence of another enzyme capable of catalyzing the changes in neutral sugars. They proposed an additional hydrolase(s) to explain the decrease in galactose and arabinose during ripening. The hydrolase was either absent in the extract or inactive *in vitro*. The absence of PGA activity in apples during softening reported by Knee (1974) leads one to conclude that such hydrolases are present. These may be glycosidases, several of which have been identified in apples and tomatoes by Bartley (1974) and Wallner and Walker (1975).

V. EFFECT OF CALCIUM IONS ON THE TEXTURE OF CANNED FRUITS AND VEGETABLES

The canning of fruits and vegetables is generally accompanied by a softening of the texture during processing. Kertesz (1939) observed that

TABLE 8.3

Change in Sugar Composition of Noncellulosic, Neutral Cell Wall Polysaccharides During Ripening and After *in Vitro* Enzyme Treatment[a,b]

	In situ			*In vitro*	
Sugar	Green	Turning	Red	Boiled control	Active enzyme
Rhamnose	1.45	1.68	1.60	1.41	1.45
Arabinose	3.23	2.99	2.58	3.22	3.15
Xylose	3.35	3.62	3.97	3.22	4.28
Mannose	1.75	2.12	2.28	1.80	2.25
Galactose	12.07	9.79	7.32	11.94	11.61
Glucose	2.41	2.46	2.51	2.35	3.23

[a] From Wallner and Bloom (1977).

[b] Data are expressed as milligrams sugar per 100 mg wall.

calcium salts had a firming effect on plant tissues and subsequently developed a method for treating tomatoes with calcium chloride before canning (Kertesz, 1939; Kertesz *et al.*, 1940). Loconti and Kertesz (1941) confirmed this effect by showing that the addition of soluble calcium salts to tomatoes before canning produced a firmer product. This firming effect appears to be due to the formation of insoluble calcium pectates by calcium ions cross-linking adjacent pectic acid chains through ionic bonding with the free carboxyl groups, as shown below. These bridges prevent solubilization of the pectic substances and result in a firmer product.

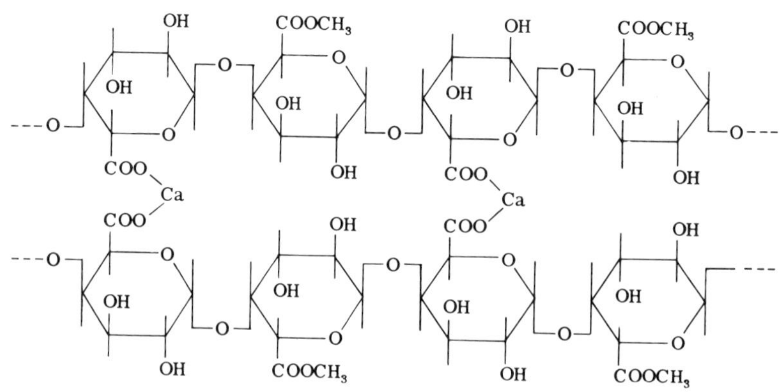

TABLE 8.4

Effects of Blanching Time and Temperature on the Firmness of Golden Delicious Apple Slices[a,b]

Treatment	Blanching condition					
	20 seconds (71°C)		60 seconds (82°C)		120 seconds (99°C)	
	After blanching	After canning	After blanching	After canning	After blanching	After canning
Slice + H_2O	50a	50a	50a	60a	50a	60a
Slice + H_2O + Ca	72b	77b	70b	70b	65a	65a
Slice + H_2O + PE	55a	52a	55a	60a	50a	50a
Slice + H_2O + PME + Ca	80b	75b	72b	70b	60a	60a

[a] From Wiley and Lee (1970). Copyright by Institute for Food Technologists.

[b] The data are shear press readings in pound-force. The firmness of raw slices was 250 lb-force. Means (average of two observations) not followed by the same letter are significantly different at the 5% level; $r = 0.98$ between duplicates.

A similar improvement in the texture of thermally processed apple slices following the addition of calcium salts was also reported by Collins and Wiley (1963). A further attempt was made by Wiley and Lee (1970) to improve the texture of Golden Delicious apple slices by treatment of the slices with calcium chloride, pectinesterase, and a combination of the two. The results (Table 8.4) indicate clearly that the apple slices processed with both calcium chloride and pectinesterase were considerably firmer than those treated with only calcium chloride or pectinesterase. The effect of calcium chloride and pectinesterase on firmness was reduced as the blanching temperature was increased.

Mohammadzadeh-Khayat and Luh (1968) added calcium chloride at 0, 50, 100, and 200 ppm to syrup before canning apricots and monitored the firmness of the canned fruit using the Lee Kramer Shear Press. The results (Fig. 8.7) show that apricots treated with 100 ppm calcium chloride peaked in firmness and relative to the apricots treated with smaller amounts of calcium chloride led in firmness throughout the 9-month storage period. The sample treated with 100 ppm calcium

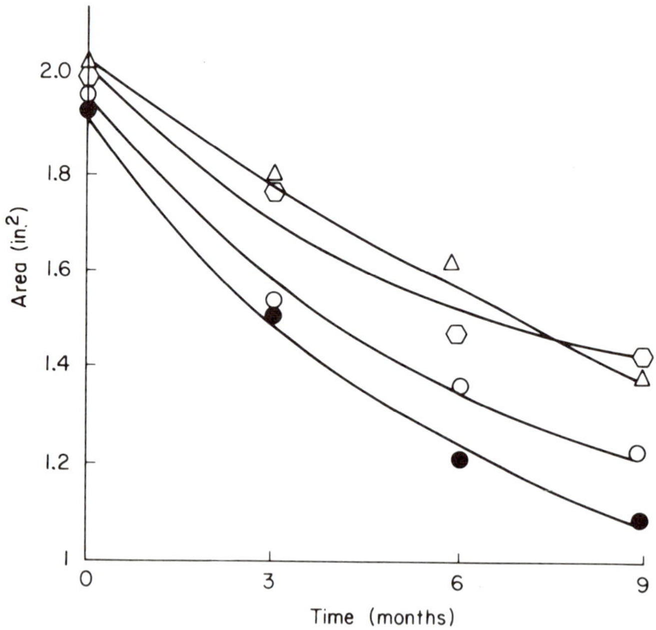

Fig. 8.7. Effect of storage (21°C) on the texture of canned apricots. Syrups treated with 0 (●), 50 (○), 100 (△), and 200 (⬡) ppm calcium (Mohammadzadeh-Khayat and Luh, 1968).

chloride was as firm as the sample treated with 200 ppm calcium chloride. Van Buren (1968) used additions of calcium chloride to produce firmer canned string beans. Firmness was most pronounced when calcium chloride was added before blanching. These studies all show the effectiveness of calcium ions in improving the texture of canned fruits and vegetables.

Chapter 9

Lignin

I. INTRODUCTION

Lignin is a complex polymer of phenylpropanoid (C_6–C_3) units encrusting and penetrating the cell walls of tracheid vessels, fibers, and sclereids of vascular plants (Freudenberg and Neish, 1968). It plays an important role in the internal transport of water, nutrients, and metabolites by decreasing the permeation of water across the cell walls in the conducting xylem tissues. Lignified tissue is particularly effective in resisting fungal attack and appears to play an active role in preventing fungal penetration of the epidermal cell walls (Hijwegen, 1963; Ride, 1975; Vance and Sherwood, 1976). Lignin also provides mechanical strength for the plant cell walls and contributes to the texture of a variety of plant foods. In celery, for example, the distinctive strings are composed of highly lignified vascular tissue. In asparagus spears, excessive amounts of lignification of the fibrovascular bundles causes toughness (Herner, 1973; Segerlind and Herner, 1972). The development of sclereids, lignocellulose complexes causing grittiness in some varieties of pears, is another example of how lignification can adversely affect the textural quality of edible plants (Ranadive and Haard, 1972b).

II. CHEMICAL NATURE OF LIGNIN

Freudenberg and Neish (1968) described lignin as an insoluble constituent of the cell wall derived from the enzymatic dehydrogenation and polymerization of coumaryl, coniferyl, and sinapyl alcohols. Coniferyl

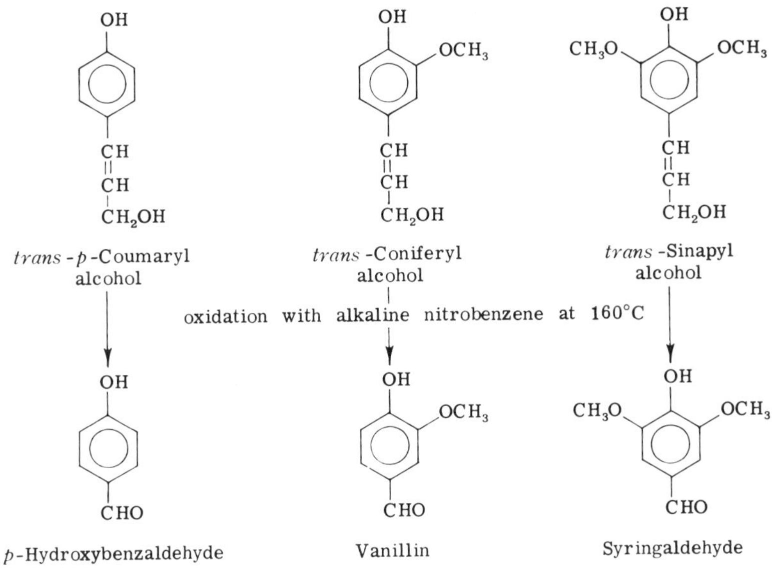

	p-Hydroxybenzaldehyde	Vanillin	Syringaldehyde
Gymnosperms	–	+	–
Monocotyledons	+	+	–
Dicotyledons	–	+	+

Scheme 9.1. The lignins of gymnosperms, monocotyledons, and dicotyledons (+, present; –, absent) (Walker, 1975).

alcohol, the basic unit of all lignins, is found in all plants. The presence of sinapyl and *p*-coumaryl alcohols varies with the plant source. For example, *p*-coumaryl alcohol is present in monocotyledons, whereas sinapyl alcohol is present in dicotyledons. In Scheme 9.1 it is shown that different aromatic aldehydes are produced when lignins from gymnosperms, monocotyledons, and dicotyledons are subjected to oxidation in the presence of alkaline nitrobenzene.

III. LIGNIN BIOSYNTHESIS

Lignin is synthesized by the plant from carbohydrates by way of the shikimic acid pathway, as outlined in Scheme 9.2. A series of tracer studies showed that both L-phenylalanine and cinnamic acid were effective presursors of lignin in mono- and dicotyledons (Brown and Neish,

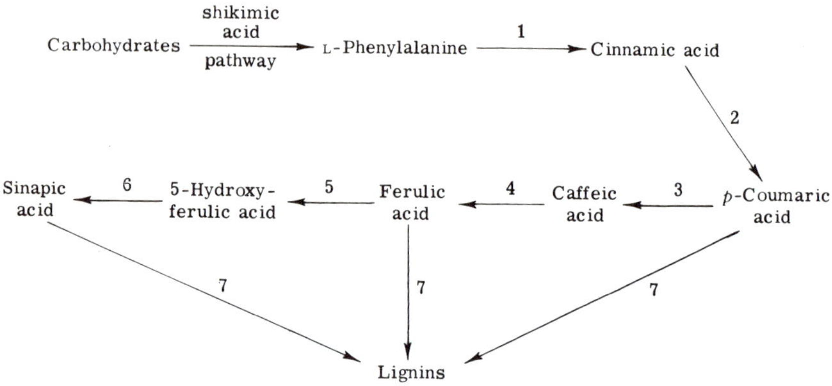

Scheme 9.2. Biosynthesis of lignins. Steps 1–6 are catalyed by five enzymes: 1, phenylalanine ammonia lyase; 2, cinnamate 4-hydroxylase; 3, 4-hydroxycinnamate 3-hydroxylase; 4 and 6, *O*-methyltransferase; 5, possibly ferulate 5-hydroxylase (2, 3, and 5 are probably all mixed-function oxidases). Step 7 includes CoA ligases to form CoA esters, which are then reduced to alcohols and oxidatively polymerized to lignin in the cell wall by peroxidases. (Adapted from Stafford, 1974.)

1955a,b, 1956, 1959; Wright *et al.*, 1958; McCalla and Neish, 1959. Most of the research on lignin has been related to the development of wood; however, wherever possible reference will be made to edible plants.

A. Phenylalanine Ammonia Lyase

Kokoul and Conn (1961) isolated an enzyme, phenylalanine ammonia lyase (PAL; EC 4.3.1.5), from barley that catalyzed the deamination of L-phenylalanine to *trans*-cinnamic acid. The formation of cinnamic acid by PAL is particularly important in plants because cinnamic acid is the presursor of a number of secondary metabolites including lignin. Phenylalanine ammonia lyase has been studied extensively in plants and has been discussed in several recent reviews (Camm and Towers, 1973a; Creasy and Zucker, 1974; Yoshida, 1969). It has been identified in sweet potato roots (Minimakawa and Uritani, 1965), mung beans (Ahmed and Swain, 1970), buckwheat (Amrhein and Zenk, 1971), mustard plants (Schopfer, 1971), swedes and parsnips (Rhodes and Wooltorton, 1971), and carrots (Sugano *et al.*, 1975). Although PAL activity has been generally associated with the soluble fraction, recent evidence supports the localization of this enzyme in microsomes, chloroplasts, and microbodies (Ruis and Kindl, 1970, 1971; Monties, 1974; Weissenbock, 1975, Gregor, 1976).

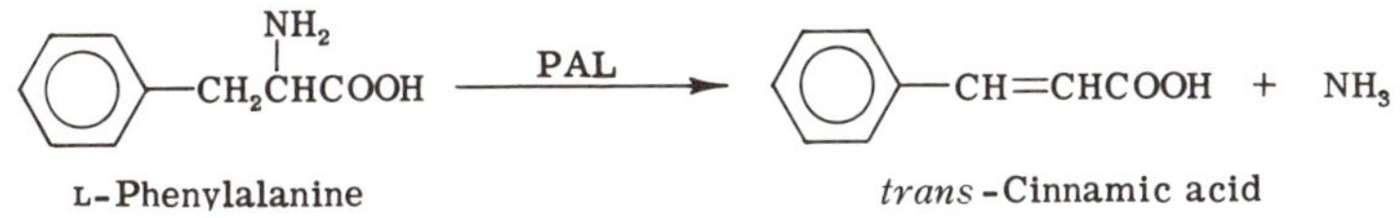

In contrast to dicotyledons, monocotyledons are capable of converting L-tyrosine to lignin. This conversion occurs in grasses through deamination of L-tyrosine to *p*-coumaric acid by an enzyme, tyrosine ammonia lyase (TAL) (Neish, 1961).

B. Cinnamate 4-Hydroxylase

The next step in lignin biosynthesis is the hydroxylation of cinnamic acid by cinnamate 4-hydroxylase (Neish, 1968). The presence of this enzyme in the microsomal fraction of pea seedlings was first reported by Russel and Conn (1967) and was specific for hydroxylation in the para position. It has since been identified in buckwheat (Amrhein and Zenk, 1968), spinach (Nair and Vining, 1965), asparagus shoots (Shimada *et al.*, 1970a), and potato tuber (Camm and Towers, 1973b). Hill and Rhodes (1975) isolated cinnamate 4-hydroxylase from swede root disks and found that its properties were similar to those of the corresponding pea seed enzyme. Both enzymes were located in the microsomal fraction and had a specific requirement for NADPH and cinnamic acid.

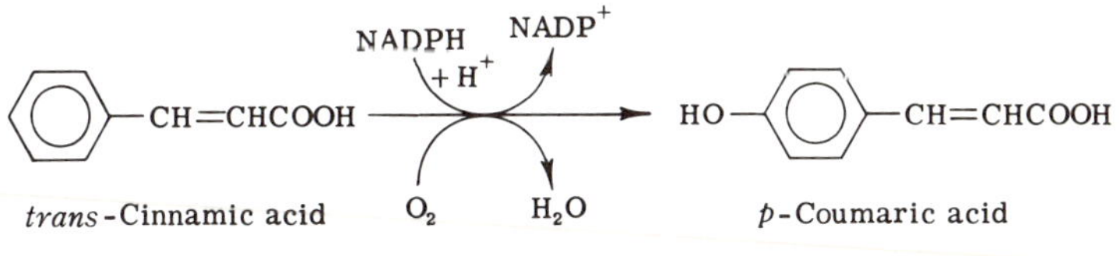

Cinnamate 4-hydroxylase is a monooxygenase or mixed-function oxidase requiring NADPH and molecular oxygen. Studies with animal systems indicate similar oxidases in which reduced cytochrome *P*-450 is involved in binding or activating molecular oxygen. The involvement of cytochrome *P*-450 in the hydroxylation of cinnamic acid by pea seedlings was reported by Russel (1971). This was confirmed by Potts *et al.* (1974), who found that the CO difference spectrum and photodissociation of the CO complex of *Sorghum* cinnamate 4-hydroxylase were identical to those of the cytochrome *P*-450 hydroxylase system.

The close relationship between PAL and cinnamate 4-hydroxylase was identified in cucumber cotyledons by Czicki and Kindl (1977). These researchers demonstrated the presence of both these enzymes in the microsomal membrane sequentially catalyzing the formation of coumaric acid.

C. 4-Hydroxycinnamate Hydroxylase

The enzymatic hydroxylation of *p*-coumarate to caffeic acid is not clear, although it appears to be associated with diphenolase activity. Several phenol oxidases were identified in sorghum by Stafford and Dresler (1972). One, which was extracted from the first internode, was capable of hydroxylating *p*-coumarate. It was located in the soluble fraction together with PAL and TAL. This contrasted with the earlier results of Stafford and Baldy (1970), who identified a similar enzyme complex in the particulate fraction of sorghum that was also capable of converting phenylalanine to caffeic acid. Stafford (1975) reported that 4-hydroxycinnamate hydroxylase could be released from the microsomal fraction in a lower molecular weight form following treatment with the sulfhydryl reagents 2-mercaptoethanol and dithiothreitol. The enzyme was partially characterized by Stafford (1976) and appeared to require ascorbate or reduced pyridine nucleotide as electron donor. Parish (1972) reported *p*-coumarate hydroxylating activity in a phenol oxidase preparation from leaf extracts of *Beta vulgaris*. The activity appeared in the chloroplast and supernatant. It confirmed earlier observations by Sato (1966) that the chloroplasts of gymnosperms and angiosperms exhibited similar enzyme activity.

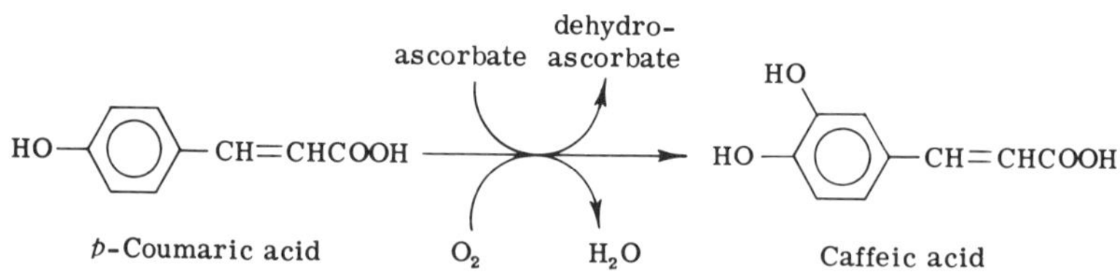

The next hydroxylation step appears to be the conversion of 3-methoxy-4-hydroxycinnamate (ferulic acid) to 3-methoxy-4,5-dihydroxycinnamate (5-hydroxyferulic acid). The biochemistry of this reaction remains to be elucidated, although it appears to be an important step in the formation of sinapic acid.

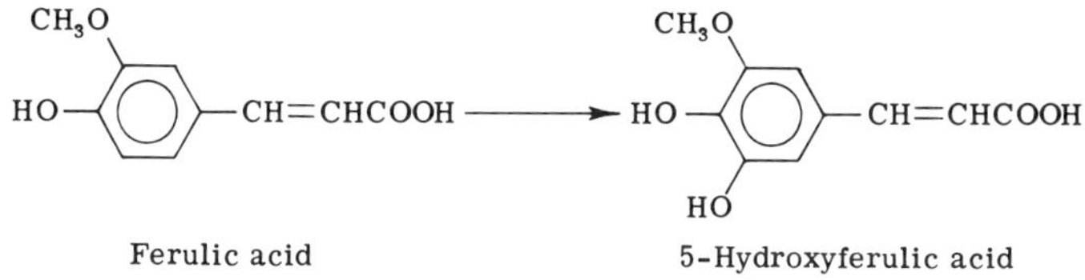

D. *O*-Methyltransferases

After hydroxylation, methylation of caffeic and 5-hydroxyferulic acids to ferulic and sinapic acids, respectively, probably occurs (Scheme 9.3). Byerrum *et al.* (1954) demonstrated the incorporation of methyl groups from methionine into methoxyl groups of lignin in barley plants. This was catalyzed by the enzyme *S*-adenosylmethionine : catechol *O*-methyltransferase (EC 2.1.1.6), which required Mg^{2+} and *S*-adenosylmethionine (SAM) as methyl donor. This enzyme is present in animal tissue and has been identified in higher plants (Finkle and Nelson, 1963; Finkle and Masri, 1964; Hess, 1965a,b). *O*-Methyltransferase, although meta-specific, exhibits differences in substrate specificity in angiosperms and gymnosperms. The gymnosperm enzyme methylates the hydroxyl group on C-3 of caffeic acid far more readily than 5-hydroxyferulate. The enzyme from angiosperm can methylate both hydroxyls. This difference in enzyme specificity accounts for the fact that there are guaiacyl (3-methoxy-4-hydroxyphenyl) units in gymnosperm lignins and both syringyl (3,5-dimethoxy-4-hydroxyphenyl) and guaiacyl units in angiosperm lignins.

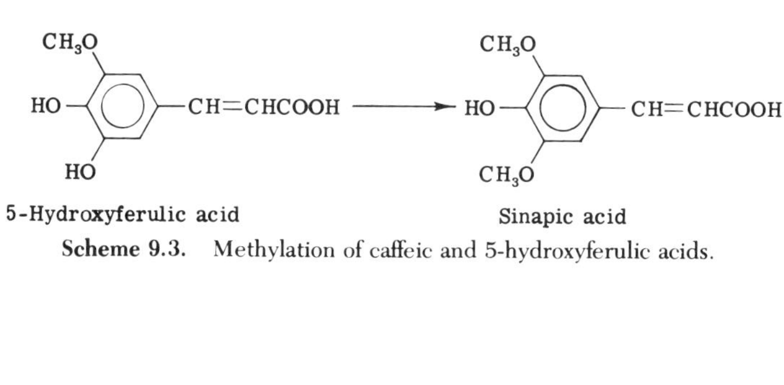

Scheme 9.3. Methylation of caffeic and 5-hydroxyferulic acids.

Shimada *et al.* (1970b) examined the methylation of various hydroxycinnamic acids by angiosperm bamboo and poplar *O*-methyltransferases and compared it with the methylation by *O*-methyltransferase of the gymnosperm *Ginka biloba*. Both bamboo and poplar enzymes readily methylated 3,4-dihydroxy-3,4,5-trihydroxy-, and 3,4-dihydroxy-5-methoxycinnamic acids, whereas only 3,4-dihydroxycinnamic acid was methylated by the *Ginka biloba* enzyme. Shimada *et al.* (1972) reported a broader specificity of the gymnosperms pine (*Pinus thumbergii*) *O*-methyltransferase. Purification of pine *O*-methyltransferase by Kuroda *et al.* (1975) showed that it was eluted off an ion-exchange column by a more dilute KCl solution than the corresponding bamboo enzyme. Differences in elution pattern and a slower mobility of the pine enzyme on polyacrylamide gel electrophoresis, due to a lower negative charge, were attributed to the difference in specificity of the pine and bamboo enzymes.

Rhodes *et al.* (1976) isolated *O*-methyltransferase from swede root disks. The enzymes was specific for methylation in the meta position of caffeic acid, although it could also methylate 5-hydroxyferulic acid but at a much slower rate.

E. Reduction of Hydroxycinnamic Acids

The incorporation of hydroxycinnamic acids into lignin appears to involve reduction to the corresponding aldehyde and alcohol. Higuchi and Brown (1963) reported that labeled coniferyl aldehyde and coniferyl alcohol were formed when [^{14}C]-ferulic acid was incubated with cambial tissue of *Pinus strobus* and stems of wheat plants. The specific activity of the aldehyde was found by Brown (1966) to be higher than that of the corresponding alcohol, although the reduction of ferulic acid to the corresponding aldehyde could not be confirmed in later studies (Brown, 1969; Zenk and Gross, 1974). Nevertheless, research by Mansell *et al.* (1972) with cell-free systems from the cambial tissue of *Salix alba* demonstrated for the first time that higher plants could reduce ferulic acid to coniferyl alcohol. The reaction, requiring ATP, CoA, and reduced pyridine nucleotides suggested that feruloyl-CoA and coniferyl aldehyde were formed as intermediates. Gross *et al.* (1973) isolated three distinct enzymes from the cambial tissue of *Salix alba* (Scheme 9.4). These were identified as ferulate-CoA ligase, feruloyl-CoA reductase, and coniferyl alcohol oxidoreductase, which catalyze the reaction sequence leading to coniferyl alcohol (Scheme 9.4).

Several researchers (Hahlbrock and Grisebach, 1970; Walton and Butt,

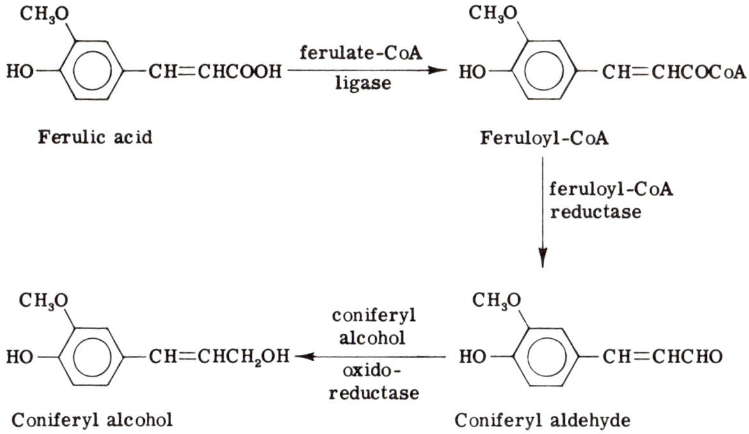

Scheme 9.4. Reduction of ferulic acid to coniferyl alcohol.

1971) described enzymes in higher plants that were capable of activating cinnamic acids in the presence of ATP and CoA to form CoA esters. Rhodes and Wooltorton (1973) identified an enzyme in swede root disks that was responsible for the formation of CoA thio esters of cinnamic acid in the presence of CoA, ATP, and Mg^{2+}. The enzyme was active only on cinnamic acid derivatives possessing a phenolic hydroxyl group such as *p*-coumaric and ferulic acids. It differed in its specificity from a similar enzyme identified in spinach beet by Walton and Butt (1970. 1971). The spinach beet enzyme was most active on cinnamic acid, but there was little activity on the corresponding hydroxycinnamic derivatives. The formation of CoA thio esters is necessary because the conversion of cin namic acid to aldehyde is thermodynamically unfavorable.

Rhodes and Wooltorton (1974) isolated an enzyme system in swede root tissue that reduced cinnamic acid CoA thio esters. The enzyme system in aged swede root disks was stimulated by Mg^{2+} and required a heat-labile factor, CoA, ATP, and $NADPH_2$ to reduce *p*-coumarate to *p*-coumaryl alcohol. The requirement for CoA, ATP, and Mg^{2+} was similar to the requirement of *p*-coumaryl-CoA synthetase reported earlier by Rhodes and Wooltorton (1973). A further enzyme was found that required $NADPH_2$ or $NADH_2$ depending on the nature of the substrate. For example, if *p*-coumarate was substrate, $NADPH_2$ was required. In the presence of ferulic acid either $NADPH_2$ or $NADH_2$ could catalyze the reaction. Rhodes and Wooltorton (1975) characterized the enzyme system involved in the reduction of ferulic acid to coniferyl alcohol by the use of DEAE-cellulose. The enzymes included hydroxycinnamate-CoA ligase, feruloyl-CoA reductase, and an aromatic alcohol dehydrogenase. Feruloyl-CoA reductase was further separated into two fractions, each requiring $NADPH_2$ and exhibiting a high affinity for the substrate. The aromatic alcohol dehydrogenase was also resolved into two distinct fractions. The second fraction was partially separated into two subfractions from the extracts of aged swede root disks. Both subfractions required NADP, although the second fraction showed a greater affinity for the substrate. Feruloyl-CoA reductase was present in the initial disks and increased only twofold during aging. The activity of the alcohol dehydrogenase fractions remained constant with some qualitative changes in fraction 2. In contrast, hydroxycinnamate CoA-ligase, although present at very low levels initially, increased markedly during aging of the disks. This suggested that the reducing system in aged swede root disks was controlled by hydroxycinnamate-CoA ligase.

F. Polymerization

Polymerization of alcohols such as coniferyl alcohol represents the final stage of lignification. The conversion of these phenylpropanoid (C_6–C_3) precursors to an insoluble ligninlike material, referred to as dehydrogenative polymerization product (DHP), was demonstrated by *in vitro* studies. Freudenberg *et al.* (1952) reported polymerization of coniferyl alcohol into this ligninlike polymer in the presence of a crude mushroom polyphenol oxidase preparation. Higuchi (1957) suggested that the enzyme described by Freudenberg *et al.* (1952) was probably laccase (*p*-diphenol: O_2 oxidoreductase, EC 1.10.3.2). He examined mushroom polyphenol oxidase and reported the presence of both tyrosinase and laccase (Higuchi, 1958). Mason and Cronyn (1955) also investigated the mushroom preparation and suggested that coniferyl alcohol was oxidized by a heat-labile factor other than polyphenol oxidase. Siegel (1953) reported the conversion of euphenol to DHP in sliced embryonic hypocotyls of red kidney beans in the presence of hydrogen peroxide. A series of studies on lignin formation in bamboo shoot by Higuchi (1958) and Higuchi *et al.* (1954, 1955) also confirmed the importance of peroxidase (EC 1.11.1.7). Nakamura (1967) separated three enzymes from the latex of Japanese lacquer tree (*Rhus vermicifera*): a phenolase, peroxidase, and laccase. Of these enzymes only peroxidase catalyzed the oxidation of coniferyl alcohol. The role of peroxidase was also established by Nozu (1968), who incubated coniferyl alcohol with a purified peroxidase preparation from bamboo shoot and obtained a DHP closely resembling native lignin except for its lower molecular weight. Dehydrogenative polymerization product was also formed with wheat germ and horseradish peroxidases, which confirmed the participation of plant peroxidases in lignin formation.

The exclusive participation of peroxidase in lignification was demonstrated by Harkin and Obst (1973) in histochemical studies on cross sections of tree branches, saplings, and cambial tissue cultures. These researchers used (a) syringaldazine, a pale yellow crystalline compound readily oxidized by laccase or peroxidase to a conjugated compound, (b) tetramethoxyazo-*p*-methylenequinone, which has an intensely purple color (Scheme 9.5). This color change was observed only in cross sections where hydrogen peroxide had been added, providing further evidence for the sole participation of peroxidase in the polymerization of alcohols to lignin.

G. Peroxidase Mechanism and Localization

The role of peroxidase in lignin formation appears to be related to its ability to generate free radicals in phenylpropanoid units, such as coniferyl alcohol (Scheme 9.6). These then condense to form polymers (Freudenberg, 1968). The presence of peroxidase was reported in lignified secondary cell walls of the tracheary tissue in bamboo shoots (Higuchi, 1957), sunflower crown gall tissue, and normal and diseased elm trees (Lipetz and Garro, 1965). A number of researchers, however, were unable to demonstrate the presence of peroxidase in the xylem wall during lignification. Studies by Wardrop and Bland (1958) on *Eucalyptus* and *Pinus* and by Koblitz (1961) on *Taxus*, *Populus*, and *Aesculus* found that peroxidase activity in young xylem elements disappeared once lignification occurred. Later studies on root tips of *Allium* by DeJong (1967) and woody tissue of *Robinia* and *Acer* by Czaninski and Catesson (1969) did not establish peroxidase activity in the lignified walls of tracheary tissue. Hepler *et al.* (1972) attempted to resolve this controversy by examining

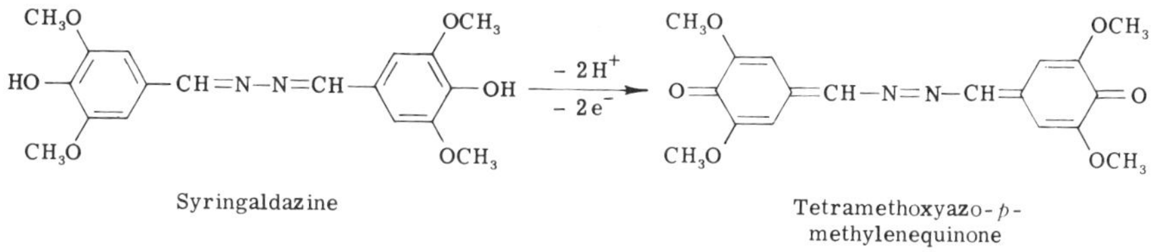

Scheme 9.5. Oxidation of syringaldazine to tetramethoxyazo-*p*-methylenequinone (Harkin and Obst, 1973).

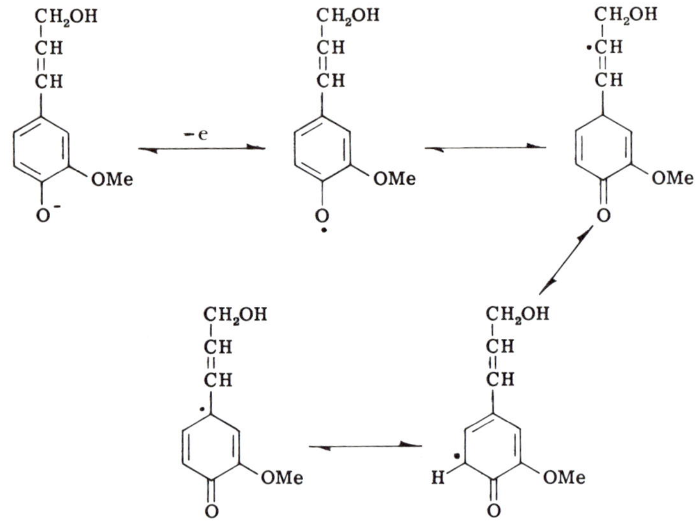

Scheme 9.6. Production of free radicals during dehydrogenation of coniferyl alcohol (Freudenberg, 1968).

the wound vessels of *Coleus* by electron microscopy using the peroxidase-catalyzed oxidation of 3,3′-diamino-benzidine (DAB). With the aid of the hydrogen peroxide staining technique of Graham and Karnovsky (1966), they demonstrated the presence of peroxidase in the secondary cell walls and lignified portions of the primary cell walls. The presence of peroxidase was confirmed by the reduction of staining in the absence of hydrogen peroxide or by the addition of KCN (a known inhibitor of heme-containing enzymes). The possibility that catalase, another heme-containing enzyme, played a role in lignification was eliminated when the addition of 3-amino-1,2,4-triazole, a known inhibitor of this enzyme, did not affect the staining procedure. Cytochrome oxidase, another heme-containing protein, was ruled out by the absence of DAB staining in the mitochondria. These results showed that the localized stain in *Coleus* wound xylem elements was due to peroxidase.

Hepler *et al.* (1972) observed peroxidase accumulation in the lignified areas of *Coleus* cell walls, in the secondary wall, and in corners of both vessel and parenchyma tissue. These researchers suggested that peroxidase was localized in the dicotysomes, as observed in root meristems of *Cucium* by Poux (1969). Peroxidase appears to be synthesized in cells undergoing differentiation and then migrates to the bands of secondary cell walls via cytoplasmic microtubules (Hepler and Fosket, 1971). As vesicle fusion takes place, peroxidase probably binds to the vascular membrane and becomes part of the plasmalemma in localities associated with secondary thickening. Peroxidase then diffuses from the plasmalemma into the primary and secondary walls and catalyzes lignification.

IV. ENZYMATIC CONTROL OF LIGNIFICATION IN SWEDE ROOT DISKS

Rhodes *et al.* (1976) monitored the activity of nine enzymes in lignin synthesis by aging swede root disks of which four are shown in Fig. 9.1. They found that peroxidase, aromatic alcohol dehydrogenase, and phenylalanine transaminase underwent little change during the aging

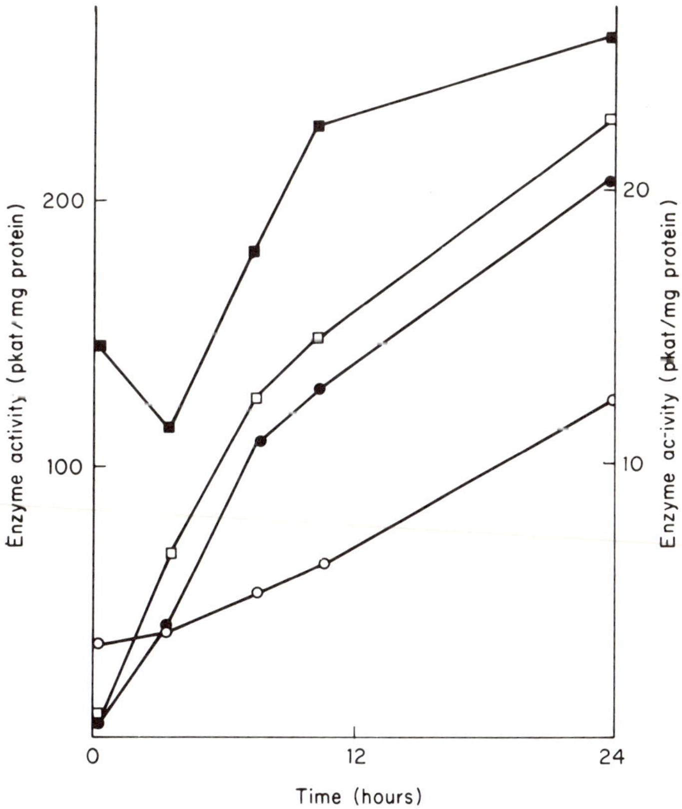

Fig. 9.1. Changes in enzyme activity during the aging of swede root disks in the presence of 8 ppm ethylene. Key: ■, shikimate dehydrogenase; □, phenylalanine ammonia lyase; ●, cinnamate 4-hydroxylase; ○, *O*-methyltransferase (Rhodes *et al.*, 1976).

period. *O*-Methyltransferase, shikimate dehydrogenase, and feruloyl-CoA reductase increased only two- to threefold and were insensitive to ethylene. The activity of phenylalanine ammonia lyase, cinnamate-4-hydroxylase, and hydroxycinnamate-CoA ligase, however, all increased around 20- to 30-fold. Addition of ethylene to these enzymes increased their activity a further 2- to 5-fold. These results suggested that in swedes, as in other plant tissues, PAL, cinnamate-4-hydroxylase, and *p*-coumarate-CoA ligase play a central role in regulating the biosynthesis of phenylpropanoid compounds in lignin formation.

V. PENTOSE PHOSPHATE PATHWAY AND LIGNIFICATION

An increase in the activity of the pentose phosphate pathway was reported to accompany lignification in explants of *Coleus blumei* and storage tissue of Jerusalem artichokes (Pryke and ap Rees, 1976) and pea roots (Wong and ap Rees, 1971). This suggested that the pentose phosphate pathway might provide some of the reducing power (NADPH) required for lignin biosynthesis. Pryke and ap Rees (1977) incubated [^{3}H,^{14}C]glucose with 4-day cultured explants of the stem of *Coleus blumei*, storage tissue of Jerusalem artichokes (*Helianthus tuberosus* L.), and roots of 5-day-old seedlings of *Pisum sativum*. They examined the ^{3}H/^{14}C ratios in the aromatic aldehydes of lignin produced by oxidation with alkaline nitrobenzene and also of lignin extracted under milder conditions by complexing with thioglycollic acid (Freudenberg *et al.*, 1959) or with dioxan (Bjorkmann, 1954). Significant levels of ^{3}H and ^{14}C were recovered in syringaldehyde, vanillin, *p*-hydroxbenzaldehyde, and lignothioglycollic acid from explants of *Coleus* and artichokes and in vanillin, *p*-hydroxybenzaldehyde, and milled-wood lignin (dioxan-extracted lignin) from pea roots. The similarity of ^{3}H/^{14}C ratios in the lignin fractions obtained from the three tissues suggested that these labeling patterns resulted from normal lignin synthesis. On the basis of these results, Pryke and ap Rees (1977) concluded that considerable reducing power for lignin synthesis was derived from the pentose phosphate pathway.

VI. SCLEREID FORMATION IN PEARS

The presence of sclereids or grit cells in the pulp of certain varieties of pears lowers the quality of the fruit. Crist and Batjer (1931) suggested that sclereid tissue and lignin from wood pulp were chemically similar. This was confirmed by Ranadive and Haard (1972a), who characterized the sclereid tissue as a lignocellulose complex. The development of sclereids was reported by Ryugo (1969) to be either characteristic of certain fruit species or due to adverse growing conditions. He also observed that in tissue regions of oriental pear fruit (*Pyrus serotina*) where sclereids developed there was considerable catalase and peroxidase activity. The role of peroxidase in lignin formation led Ranadive and Haard (1972b) to investigate the relationship between the enzyme and sclereid formation in Bartlett and Yuzuhada pears. Yuzuhada, an experimental variety at Rutgers University, differed from Bartlett by its characteristic sclereid development. Histochemical analysis of the pulp tissue sections revealed that peroxidase activity was prevalent in the parenchyma cells surrounding the grit cells as well as in the cell walls throughout the pulp. After extraction of the enzyme with maleate buffer, considerable activity in the pellet residues indicated the presence of bound peroxidase. An earlier

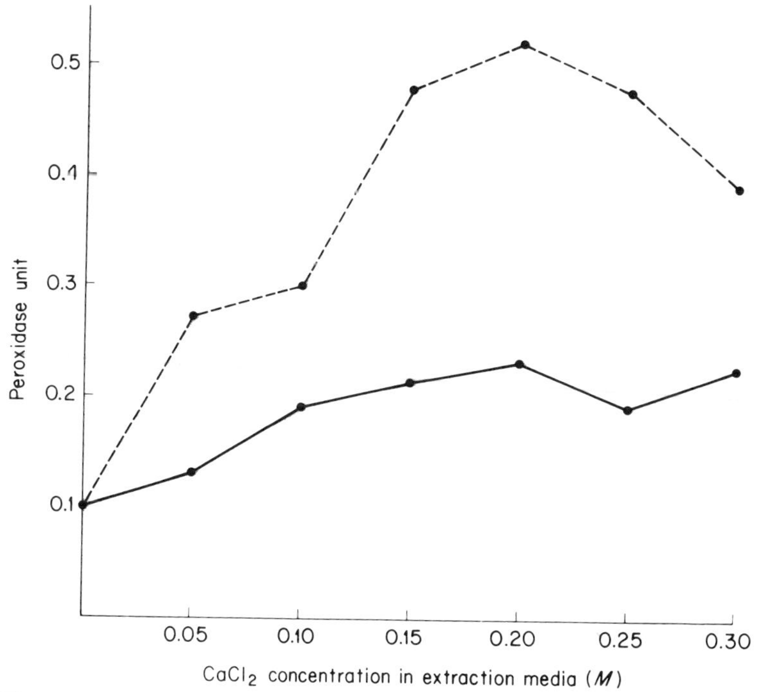

Fig. 9.2. Effect of calcium chloride on peroxidase release from pulp tissue of Bartlett (————) and Yuzuhada (----) pear fruits (Ranadive and Haard, 1972b).

TABLE 9.1

Bound Peroxidase in Developing Pear Fruit[a]

Days before harvest	Peroxidase activity[b]	
	Bartlett	Yuzuhada
21	0.19	0.43
14	0.09	0.23
7	0.02	0.26
0	0.00	0.23

[a] From Ranadive and Haard (1972b). Copyright by Institute of Food Technologists.
[b] ΔOD at 460 nm/min/mg N at 25°C.

study by Whitmore (1971) showed that bound peroxidase could be released from wheat coleoptiles by treatment with calcium nitrate. Using calcium chloride to release bound peroxidase, Ranadive and Haard (1972b) observed a marked increase in the amount of peroxidase extracted from the pulp, reaching a maximum in the presence of 0.2 *M* $CaCl_2$ (Fig. 9.2). This represented the total peroxidase fraction. The soluble peroxidase fraction was obtained by extraction in the absence of any calcium salt. The activities of both the total and soluble peroxidase fractions were consistently higher in the developing Bartlett pears than in the Yuzuhada fruit. Determination of bound peroxidase, as the difference between total and soluble peroxidase activity, revealed the opposite situation. This is evident in Table 9.1, which shows that a larger proportion of peroxidase was wall bound in the Yuzuhada fruit than in the Bartlett pears. The lower calcium content in the pulp of Yuzuhada pears during the early period of maturity was consistent with an increase in wall-bound peroxidase. Ranadive and Haard (1972b) suggested that mineral nutrition (Ca^{2+} and possibly other ions) limits peroxidase localization and thus controls the amount of lignin deposition in pears.

VII. LIGNIFICATION AND THE QUALITY OF ASPARAGUS

Toughness in asparagus is caused by lignification of the fibrovascular bundles (Bittings, 1917; Isherwood, 1955). Thus, a fiber content greater than 0.25% of the fresh weight in the asparagus spear makes it unacceptable as a food (Herner, 1973; Segerlind and Herner, 1972). Sosa-Coronel *et al.* (1976) commented that asparagus is generally harvested by snapping. This involves bending the spear and breaking it at a point where the fiber level is within acceptable limits. This method is particularly effective since the fiber content of asparagus spears increases basipetally, with the basal part being higher in fiber content (Narayan *et al.*, 1969; Smith and Kramer, 1947). Consequently, when this method of harvesting is used, 2–3 cm of spear high in fiber content are left in the field.

Clore *et al.* (1976) reported a close relationship between objectionable fiber content and maximal press peak force values for raw asparagus. These results confirmed similar findings by Kramer *et al.* (1949) and Wiley *et al.* (1956). The smaller-diameter spears were much higher in fiber content, with objectionable fiber levels closer to the tip, compared to the large-diameter asparagus spears. This indicated that the larger spears are more tender than the smaller spears.

A. Temperature and Fiber Content

The temperature during postharvest storage was found by Clore *et al.* (1976) to affect the maximal peak force of asparagus. Asparagus stored at 0°C had lower maximal values than asparagus stored at 16°–27°C. A temperature effect also occurred during the preharvest period as a result of season and weather. In this case higher maximal peak force values were observed in smaller-diameter asparagus spears (1.0–1.6 cm) following a cold period (0°C or less in the morning) compared to those harvested following a warm growing period. A similar temperature effect was reported by Sosa-Coronel *et al.* (1976), who found a lower fiber content in Mary Washington asparagus spears grown at 21.1°–26.0°C than in those grown at 10°–15°C. An increase in temperature was thought to cause spears to grow faster, resulting in lower fiber levels. Both studies confirmed observations by Kaufmann (1967) that raising the air temperature, heating the soil, or covering the ground with plastic increased the rate of spear growth and reduced the fiber content. Clore *et al.* (1976) suggested that the problem of high-fiber asparagus harvested during cold weather could be minimized by processing spears of shorter length with spears of larger diameter for packs requiring minimal amounts of fiber.

B. Distribution of Fiber in Asparagus Spears

The distribution of fiber in the top 10-cm section and four successive 2.5-cm sections of asparagus spears was examined by Sosa-Coronel *et al.*

TABLE 9.2

Percent Fiber in Spear Sections of Asparagus Cultivars[a,b]

Cultivar	Distance from the tip (cm)					Mean total fiber spear
	0.0–10.0	10.0–12.5	12.5–15.0	15.0–17.5	17.5–20.0	
U.C. 66	0.102 a[c]	0.149 a	0.165 a	0.231 a[d]	*0.605* a	0.224 a
U.C. 72	0.128 a	0.159 a	0.164 a	0.190 a	*0.502* a	0.217 a
U.C. 309	0.121 a	0.195 a	0.170 a	*0.276* a	*0.694* a	0.266 a
U.C. 711	0.163 a	0.218 a	0.211 a	*0.360* a	*0.804* a	0.280 a
Mary Washington	0.108 a	0.135 a	0.163 a	*0.320* a	*0.805* a	0.257 a
MSU-1	0.079 a	0.127 a	0.126 a	*0.259* a	*0.563* a	0.213 a
N.J. 44 × 22	0.145 a	0.216 a	0.182 a	*0.288* a	*0.658* a	0.278 a
N.J. 51 × 22	0.091 a	0.134 a	0.138 a	*0.292* a	*0.493* a	0.213 a
N. J. Improved	0.123 a	0.157 a	0.143 a	*0.378* a	*0.697* a	0.271 a

[a] Adapted from Sosa-Coronel *et al.* (1976).
[b] The spears were harvested on June 25 and were cut at ground level. Italic numbers indicate inedible spear sections.
[c] Mean separation for spear sections by Duncan's multiple range test, 5% level.
[d] The statistical analysis was performed separately for each section of the spear.

(1976) in nine cultivars. Their results (Table 9.2) show that the amount of fiber within the spears varied considerably among the cultivars studied. They suggested that careful selection of cultivars on the basis of larger amounts of usable spear could increase the edible tonnage of asparagus per acre. The higher fiber content in the basal section of the spears was consistent with observations by Segerlind and Herner (1972) that spears harvested too close to the ground were higher in fiber content. Sosa-Coronel *et al.* (1976) cautioned against the use of 0.25% fiber content to assess spear acceptability since asparagus spears with less than 0.25% total fiber had sections that were inedible due to higher fiber levels. This is illustrated in Table 9.2 for asparagus cultivars MSU-1, NJ 51 × 22, U.C. 66, and U.S. 72, which are all within the acceptable fiber limit, although the fiber content of some of the 15- to 20-cm sections was considerably higher than 0.25%. These researchers emphasized the importance of knowing the fiber distribution within the spear as well as the fiber content of the cultivars. Such knowledge would permit the grower and processor to produce a product of higher quality.

Chapter 10

Proteins

I. INTRODUCTION

The primary cell walls of higher plants all contain a hydroxyproline-rich protein called extensin (Lamport, 1962, 1965, 1967, 1970; Lamport and Northcote, 1960; Keegstra *et al.*, 1973). Originally identified by Lamport and Northcote (1960) in the cell walls of sycamore (*Acer pseudoplatanus* L.) and beans (*Phaseolus vulgaris*), extensin accounted for practically all the hydroxyproline in the cell. The name extensin was given to this protein because it was thought to play a role in the extension of the cell wall during growth (Lamport, 1965). The association of the hydroxyproline-rich protein with the primary cell walls was observed by Brown (1964) and Edelman and Hall (1964). X-ray studies by Van Etten and co-workers (1961) confirmed the structure of the bound hydroxyproline in the cell wall to be that of *trans*-4-hydroxy-L-proline:

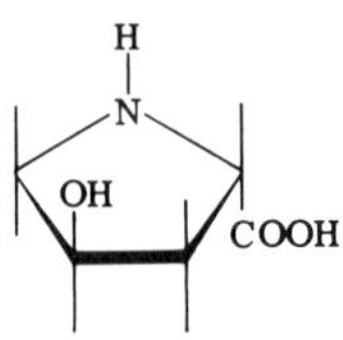

II. STRUCTURE OF EXTENSIN

A number of glycopeptides containing hydroxyproline were isolated by Lamport (1967) from tomato cell walls. Alkaline treatment of these cell walls released 70% of the hydroxyprolines as glycosides (Lamport, 1967). One of the glycopeptides released was covalently bound to a short arabinose oligosaccharide by a glycosidic bond involving carbon 1 of arabinose and the hydroxyl group of the hydroxyproline. This supported an earlier hypothesis of Lamport (1962) that extensin served as a structural protein by cross-linking with the polysaccharides of the cell wall. Continued research by Lamport and Miller (1971) led to the isolation of a series of hydroxyproline arabinosides in which mono-, di-, tri-, and tetraarabinosyl residues were attached by glycosidic bonds to the hydroxyl groups of the amino acids. These glycopeptides also contained galactose, which Lamport (1970) had identified as being linked to the hydroxyproline arabinosides by an alkali-labile bond. The existence of such a bond was discounted, however, when Lamport (1972) isolated hydroxyproline-rich galactosyl peptides essentially free of arabinose.

In a more detailed study by Lamport *et al.* (1973) the arabinose olgiosaccharides were released from the hydroxyproline residues by refluxing the cell walls of tomato tissue cell cultures at pH 1.0 for 1 hour at 100°C. The acid-stripped walls were then digested with trypsin at pH 8.0 and the products fractionated by gel filtration on Sephadex G-25 and by ion-exchange chromatography. The glycopeptides isolated by this procedure all contained *N*-serine peptide bonds. Partial acid hydrolysis of these glycopeptides revealed that the pentapeptide serine-(hydroxyproline)$_4$ was present in all of them. Although the amino acid sequences of some of these peptides were identical, they were elctrophoretically and chromatographically distinct due to the presence of variable numbers of D-galactose residues. Lamport *et al.* (1973) suggested the structure for one of the glycopeptides (S_2A_6) shown below, in which a single galactose molecule was attached to each serine. Since the majority of the cell wall

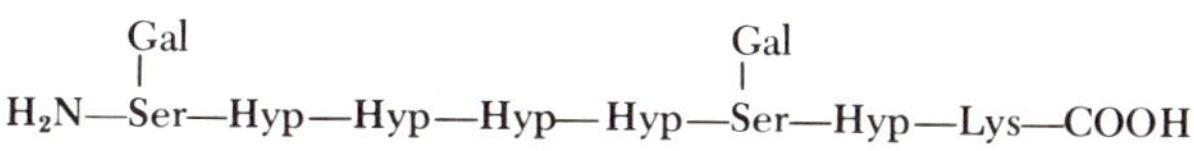

Structure of glycopeptide S_2A_6

bound hydroxyproline was *O*-substituted by tetraarabinose, these workers proposed the following structure for the wall-bound hydroxyproline:

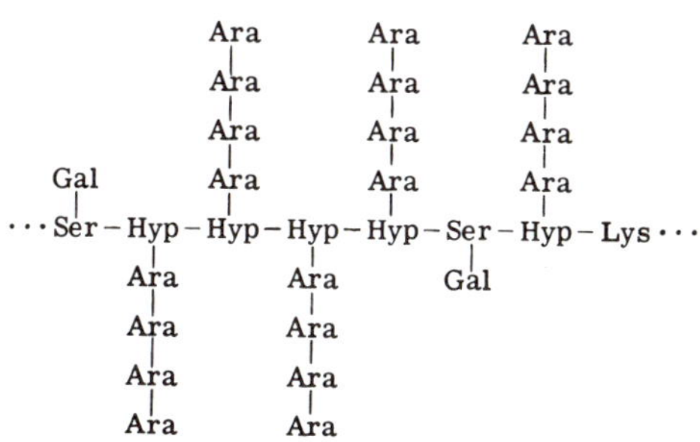

Lamport *et al.* (1973) reported that the galactosylserine bond could be cleaved by treatment with sodium hydroxide–sodium borohydride. In the light of this evidence it appeared doubtful that this bond covalently cross-linked with the major polysaccharides of the cell wall. However, the alkali-labile bond between polysaccharide and peptide components of extensin suggested by Lamport (1970) could be explained by the presence of the galactosylserine bond. The presence of this linkage in extensin from aged carrot disk cell walls was established by Cho and Chrispeels (1976). These researchers isolated the glycopeptides following acid and enzymatic treatment of the cells and separated them by gel filtration. Serine-*O*-galactosyl linkages were found in the glycopeptides but were most abundant in the hydroxyproline-rich fraction. Cho and Chrispeels (1976) suggested, however, that extensin contained a variety of short oligosaccharides attached to either hydroxyproline or serine residues. Their data did not support the presence of cross-links between extensin and the other macromolecules within the cell wall as proposed by Lamport (1965), Albersheim *et al.* (1973), and Keegstra *et al.* (1973). However, Cho and Chrispeels (1976) did not overlook the possibility that the side chains might have been larger before acid treatment of the cell walls and could not completely rule out the cross-linking role of extensin.

Extensin–Polysaccharide Network

In the extensin–polysaccharide network originally proposed by Lamport (1970), the glycoprotein was linked to cellulose microfibrils by arabinosylhydroxyproline-linked galactan. This was followed by a second model proposed by Keegstra *et al.* (1973) shown in (Fig. 7.2), in which the structural protein was covalently linked to pectic polysaccharides that were themselves covalently bound to a hemicellulose, xyloglucan. The hemicellulose fraction, xyloglucan, was reported by Bauer *et al.* (1973) to be hydrogen bonded to cellulose microfibrils. In Fig. 7.2 extensin is bound to the cell wall through a series of polymers: galactan-rhamnogalacturonan-arabinan or galactan-xyloglucan-cellulose. This model indicates that the cell wall matrix polymers are all linked by glycosidic bonds, and the sole linkage with cellulose is through hydrogen bonding by xyloglucan. Monro *et al.* (1976) studied the primary cell wall structure of lupin and mung bean hypocotyls and were unable to explain their observations in terms of the model proposed by Keegstra *et al.* (1973). Monro *et al.* (1976) indicated that polyuronide bridging was not involved in binding extensin to the cell wall nor were galactoserine linkages required to bind himicellulose to the wall. These researchers proposed a model in which a fraction of the wall protein was strongly bound to cellulose microfibrils or to other polysaccharides that were themselves linked to cellulose. It is apparent from the various studies discussed here that considerable clarification is required to establish the precise arrangement of extensin in the primary cell wall structure.

III. EXTENSIN BIOSYNTHESIS

Extensin is synthesized in the cytoplasm as a proline-rich polypeptide that undergoes hydroxylation by a soluble peptidyl proline hydroxylase (Sadava and Chrispeels, 1971). Studies by Karr (1972) using suspension-cultured sycamore cells showed that glycosylation of extensin was catalyzed by arabinosyltransferases involving uridine diphosphate (UDP)-L-arabinose. Each hydroxyproline acquires a short arabinose side chain and is then secreted into the cell wall via membranous organelles, which were thought to originate from the endoplasmic reticulum and not from the Golgi apparatus (Dashek, 1970). Gardiner and Chrispeels (1975) in studies on carrot root phloem parenchyma found that the Golgi apparatus was important in the glycosylation of glycoprotein and its subsequent secretion in the cell walls. Pope (1977) monitored the pathway of ^{14}C-labeled hydroxyproline-containing proteins from the cytoplasm to the cell wall in suspension-cultured *A. pseudoplatanus* cells. His results were in agreement with those reported by Chrispeels (1969) and Dashek (1970), namely, that [^{14}C]hydroxyproline-containing protein could pass directly from a membrane-bound compartment in the cytoplasm to the cell wall. One of three hydroxyproline-rich glycoproteins identified in the cell wall by Pope (1977) was extensin. This contained mono-, tri-, and tetraarabinoside side chains and accounted for more than 90% of the total hydroxyproline.

IV. CALLOSE FORMATION IN FRUITS

The increase in firmness associated with the rough treatment of some fruits and vegetables is due to the formation of callose. At one time callose was considered to be a polysaccharide, but recent evidence points to the glycoprotein nature of this compound.

Callose was first identified in grape vine (*Vitis vinifera*) as a polysaccharide composed of 1,3-linked β-D-glucopyranose residues (Aspinall and Kessler, 1957). This polysaccharide is not a normal component of the plant cell wall but is classified as a wound-response material that acts as a plug to minimize loss of or damage to cytoplasm or to prevent penetration by infecting organisms (Bonner and Varner, 1965; Rawlins and Takahashi, 1952). Its appearance has been reported following mechanical damage (Currier and Strugger, 1956; Currier and Webster, 1964; Dekazos and Worley, 1967; Dekazos, 1972), plasmolysis (Currier and Strugger, 1956), and viral infection (Shimomura and Dijkstra, 1975).

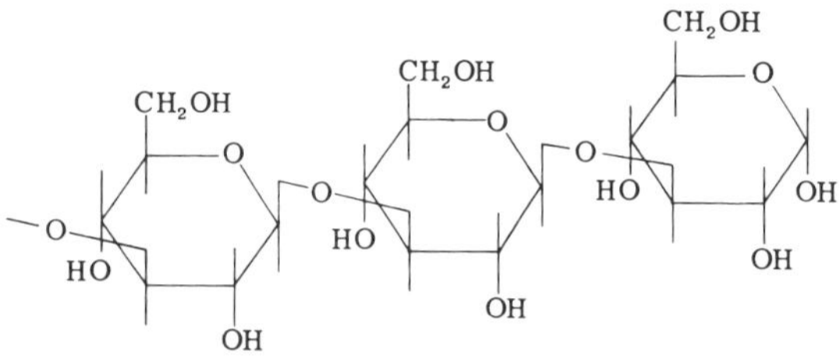

Callose or β-1, 3-glucan

Several researchers reported that, when bruised during harvesting and handling or by artificial aging, cherries became rather tough and maintained their shape after canning (Hills *et al.*, 1953; LaBelle and Moyer, 1960; Whitenberger and Hills, 1953). This phenomenon was attributed initially to changes in the pectic substances, but their removal by hot hydrochloric acid or polygalacturonase did not alter the rigidity of the cell wall (Buch *et al.*, 1961). Dekazos and Worley (1967) showed that this firming effect was due to callose formation induced by the bruising or aging of red tart cherries. Histological examination of the tissues by the use of aniline blue dye resulted in a yellow fluorescence under ultraviolet light, which increased in intensity under blue light. This is a specific test for callose formation (Currier, 1957). Callose was detected only in bruised cherry tissue, indicating that it is formed in response to injury, as observed for other plants (Currier, 1957; Evert and Derr, 1964).

The development of callose in tomatoes following bruising and heating was monitored by Dekazos (1972). Tissue sections from unbruised and unheated mature green, fresh green, red, and ripe tomatoes were found to be virtually free of callose. However, as soon as unbruised and red tomatoes were subjected to a temperature of 50°C for 1 hour, considerable callose synthesis occurred across the pericarp. This temperature dependence of callose formation in tomatoes paralleled that for cotton reported by Webster and Currier (1968), who found that a maximal amount of phloem callose was induced at 50°C. Tissue sections from artificially bruised mature green tomatoes heated for 30 minutes at 45°C showed the characteristic staining of callose formation from pit to plate as well as callose dots on the surface of the protoplast. Increasing the heating time to 1 hour caused the formation of callose deposits in the parenchyma cells. In mechanically harvested fruit, callose formation was also evident and, when the fruit was heated at 45°C for 1 hour, it increased in the parenchymatous cells until callose deposits covered the entire cell wall. In processed whole tomatoes callose was observed in the parenchymatous tissue as pit callose and heavy pit callose, covering the entire cell-wall surface. It was apparent that callose was extremely stable to heat conditions prevailing in the canning process, as reported previously by Currier (1957) and Dunning (1959). This study confirmed the relationship between callose formation in tomatoes in response to bruising and heating. Dezakos (1972) considered callose to be a major contributor to the texture of the tomato fruit as well as to the consistency of tomato products.

A. Callose Formation and Temperature

The effect of heating on the formation of callose was investigated by a number of researchers (Lerch, 1960; Webster and Currier, 1968; Dekazos, 1972). Smith and McCully (1977) examined the effect of mild temperature stress on callose synthesis in seedlings of maize (*Zea mays* L.), sorghum (*Sorghum vulgare*), peas (*Pisum sativum* L.) mung beans (*Phaseolus aureus*), soybeans (*Glycine max* L.), and tomatoes (*Lycopersicon esculentum*) using the aniline blue-induced fluorescence technique for β-1,3-glucans or callose. Large fluorescent spots were observed in the root and shoot apices of maize seedlings grown at 20°C; these were absent when the seedlings were grown at 26°C. However, no differences due to temperature were detected in callose at the level of the sieve plates and associated lateral pits of the phloem. This conflicts with studies by Webster and Currier (1965) and McNair and Currier (1968), who found that temperature-induced effects were reflected by phloem callose. Webster and Currier (1965) noted that preequilibration of seedlings in the dark before heating reduced the amount of phloem callose formed. The inability of Smith and McCulley (1977) to detect phloem callose might be related to the fact that the seedlings were grown in the dark. Positive aniline blue fluorescent deposits were observed in the cell walls of the outer root cap of soybeans grown at 20°C and to a lesser extent in mung beans, although no deposits were detected in tissues grown at 26°C. In the case of sorghum seedlings no differences were found when the seedlings were grown at either of these temperatures. These researchers suggested that the presence of large wall deposits of callose in response to low temperatures could be the first sign of some degree of stress. However, although small changes in temperature were found to exert similar histological effects on peas, mung beans, soybeans, and tomatoes, this study also showed that plant species and varieties respond differently to different temperature treatments.

B. Callose Glycoprotein

Reynolds and Dashek (1976) investigated the techniques used to specifically stain β-1,3-glucan or callose. These included light microscopy of pollen tubes stained with resorcin or resorcinol blue (lacmoid) (Kessler *et al.*, 1960; Tupy, 1959, 1960) and fluorescent microscopy of tubes suspended in aniline blue (Alves *et al.*, 1968; Preil, 1970). The specificity of the stains was checked by chemical and enzymatic extraction procedures to solubilize proteins and polysaccharides. These methods were used to evaluate the localization of callose in elongating *Lilium longiflorum* commercial variety Ace pollen tubes. Inconsistent results with the enzymatic and chemical extractions obtained using the aniline blue dye together with variable fluorescent patterns indicated that the value of this method for detecting callose was highly questionable. The results for lacmoid/resorcinol showed that removal of pectin and hemicellulose by the corresponding enzymes did not affect the lacmoid or fast green wall-staining techniques. Treatment of the tubes with exo-β-1,3-glucanase S174D caused the separation of a single-layer wall into an outer wall, which was highly irregular in shape, and an inner wall, which was clearly seen by light microscopy to be next to cytoplasm. Electron microscopy indicated two such layers encasing the nongrowing region of *L. longiflorum* pollen tubes (Dashek and Rosen, 1966; Rosen and Gawlik, 1966). Lacmoid staining of the outer layer was affected by glucanase attack, whereas the inner layer was not. This suggested that lacmoid had a partial specificity for callose. This is supported by the fact that under certain conditions KOH can degrade callose (Currier and Strugger, 1956) and eliminate tube wall affinity for lacmoid.

Proteases had a definite effect on lacmoid or fast green wall-staining methods. This was evident from a decrease in the lacmoid or fast green wall-staining patterns of tubes incubated with Pronase/pepsin. The diminished effect of the fast green wall staining was not unexpected since this dye also stains protein. However, the decrease in lacmoid wall staining caused by proteases was unexpected. A possible explanation was that lacmoid possesses an affinity for both protein and callose or that they are associated as a glycoprotein. Thus, removal of protein by proteolytic enzymes also eliminates the callosic portion. The existence of a callosic glycoprotein was first observed by Salmon (1947) for sieve tube callose. The results of incompatible crosses involving *Populus* species (Knox, 1972, 1973) and Brussels sprouts (Roggen, 1972) also suggested the possible protein nature of callose. Reynolds and Dashek (1976) confirmed that callose deposited in the pollen tube of *L. Longiflorum* was in fact a glycoprotein. The presence of a callosic glycoprotein as a component of the tube was not unexpected since the pollen tube wall was shown previously by Dashek and Harwood (1974) to contain a hydroxyproline-containing glycoprotein. The callosic glycoprotein was found to be a constituent of the nongrowing region of pollen tube.

The studies related to callose formation were all conducted on wounded or artificially aged plants. Extensin formation was also shown to increase following excision or wounding of carrot tissues (Chrispeels *et al.*, 1974) and, like callose, may be a defense mechanism against invading pathogens. It would appear that the relationship between glycoproteins, including extensin and callose formation, should be investigated further.

References

Abeles, F. B. (1969). Abscission: Role of cellulase. *Plant Physiol.* **44,** 447.

Ahmed, S. I., and Swain, T. (1970). The effect of light on the activity of enzymes of the aromatic pathway in peas and mungbeans. *Phytochemistry* **9,** 2287.

Albach, R. F., Juarez, A. T., and Lime, B. J. (1969). Time of naringin production in grapefruit. *J. Am. Soc. Hortic. Sci.* **94,** 605.

Albersheim, P., Bauer, W. D., Keegstra, K., and Talmadge, K. W. (1973). The structure of the wall of suspension-cultured sycamore cells. *In* "Biogenesis of Plant Cell Wall Polysaccharides" (F. Loewus, ed.), p. 117. Academic Press, New York.

Alves, L. M., Middleton, A. E., and Mare, D. J. (1968). Localization of callose deposits in pollen tubes of *Lilium longiflorum*. *Proc. Indiana Acad. Sci.* **77,** 144.

Amrhein, N., and Zenk, M. H. (1968). Induction of cinnamic acid 4-hydroxylase by light and wounding. *Naturwissenschaften* **55,** 394.

Amrhein, N., and Zenk, M. H. (1969). Das auftreten des NIH-shifts bei der *in vivo* und *in vitro* bildung hydroxylierter zimtsaurederivative in buchweizen hypokotylen (*Fagopgrum esculentum*). *Phytochemistry* **8,** 107.

Amrhein, N., and Zenk, M. H. (1971). Untersuchungen zur Rolle der Phenylalanin-Ammonia-Lyase (PAL) bei der Regulation der Flavanoidsynthese in Buchweizen (*Fagopyrum exculentum* moensch). *Z. Pflanzenphysiol.* **64,** 145.

Anonymous (1965). *Fed. Regist.* **30,** 13825.

Arigoni, D., Barton, D. H. R., Corey, E. J., Jerger, O., Caglioti, L., Dev, S., Ferrini, P. G., Glazier, E. R., Melera, A., Pradhan, S. K., Schaffner, K., Sternhell, S., Templeton, J. F., and Tobinaga, S. (1960). The constitution of limonin. *Experientia* **16,** 41.

Asen, S., Norris, K. H., and Stewart, R. N. (1969). Absorption spectra and color of aluminum-cyanidin-3-glucoside as influenced by pH. *Phytochemistry* **8,** 653.

Aspinall, G. O. (1970). "Polysaccharides." Pergamon, Oxford.

Aspinall, G. O., and Kessler, G. (1957). The structure of callose from the grape vine. *Chem. Ind. (London)* p. 1296.

Attaway, J. A., and Buslig, B. S. (1969). The metabolism of linalool in citrus plants. *Phytochemistry* **8,** 1671.

Attaway, J. A., Pieringer, A. P., and Barabas, L. J. (1967). The origin of citrus flavor components. III. A study of the percentage variations in peel and leaf oil terpenes during one season. *Phytochemistry* **6**, 25.

Babbitt, J. K., Powers, M. J., and Patterson, M. E. (1973). Effects of growth-regulators on cellulase, polygalacturonase, respiration, color and texture of ripening tomatoes. *J. Am. Soc. Hortic. Sci.* **98**, 77.

Bailey, S. D., Bazinet, M. L., Driscoll, J. L., and McCarthy, A. I. (1961). The volatile sulfur components of cabbage. *J. Food Sci.* **26**, 163.

Barmore, C. R., and Rouse, A. H. (1976). Pectinesterase acitivity in controlled atmosphere stored avocadoes. *J. Am. Soc. Hortic. Sci.* **101**, 294.

Bartley, I. M. (1974). β-galactosidase activity in ripening apples. *Phytochemistry* **13**, 2107.

Barton, D. H. R., Pradhan, S. K., Sternhell, S., and Templeton, J. F. (1961). Triterpenoids. Part XXV. The constitution of limonin and related bitter principles. *J. Chem. Soc.* p. 255.

Bauer, W. D., Talmadge, K. W., Keegstra, K., and Albersheim, P. (1973). The structure of plant cell walls. II. The hemicellulose of the walls of suspension-cultured sycamore cells. *Plant Physiol.* **51**, 174.

Beattie, H. G., Wheeler, K. A., and Pederson, C. S. (1943). Changes occuring in fruit juices during storage. *Food Res.* **8**, 395.

Beileski, R. L., and Laties, G. G. (1963). Turnover rates of phosphate esters in fresh and aged slices of potato tuber tissue. *Plant Physiol.* **38**, 586.

Bendix, G. H., Henry, R. E., and Strodz, N. H. (1952). Preservation of green color in canned vegetables. U.S. Patent 2,589,037.

Bennett, R. D. (1971). Acidic Limonoids of grapefruit seeds. *Phytochemistry* **10**, 3065.

Beytia, E., Valenzuela, P., and Cori, O. (1969). Terpene biosynthesis: Formation of nerol, geraniol and other phenols by an enzyme from *Pinus radiata* seedlings. *Arch. Biochem. Biophys.* **129**, 346.

Bishov, S. H., Henick, A. S., and Koch, R. B. (1960). Oxidation of fat in model systems related to dehydrated foods. *Food Res.* **25**, 174.

Bittings, K. G. (1917). The deterioration in asparagus. *Nat. Canners' Assoc. Bull.*, **11**, 18.

Bjorkmann, A. (1954). Isolation of lignin from finely divided wood with neutral solvents. *Nature (London)* **174**, 1057.

Blain, J. A. (1963). "Carotine and Carotinoide." Steinkopff, Darmstadt.

Blain, J. A., and Barr, T. (1961). Destruction of linoleate hydroperoxide by soya extracts. *Nature (London)* **190**, 538.

Blain, J. A., and Styles, E. C. C. (1959). A lipoperoxidase factor in soya *Nature* (London) **184**, 1141.

Blair, J. S. (1940a). Color stabilization of green vegetables. U.S. Patent 2,186,003.

Blair, J. S. (1940b). Color stabilization of green vegetables. U.S. Patent 2,189,774.

Bloch, K., Chaykin, S., Philips, A. H., and de Waard, A. (1959). Mevalonic acid pyrophosphate and isopentyl pyrophosphate. *J. Biol. Chem.* **234**, 2594.

Bone, D. H., and Fowden, L. (1960). Transamination and the oxidation of some amino acids by mung bean mitochondria. *J. Exp. Bot.* **11**, 104.

Bonner, J., and Varner, J. E., eds. (1965). "Plant Biochemistry," 2nd ed. Academic Press, New York.

Borchgrevinck, N. C., and Charley, H. (1966). Color of cooked carrots related to carotene content. *J. Am. Diet. Assoc.* **49**, 116.

Borodin, J. (1882). *Bot. Ztg.* **40**, 608; cited in *J. Biol. Chem.* **133**, 551 (1940).

Brady, C. J. (1976). The pectinesterase of the pulp of the banana fruit. *Aust. J. Plant Physiol.* **3**, 163.

Brewster, L. C., Hasegawa, S., and Maier, V. P. (1976). Bitterness prevention in citrus juices. Comparative activities and stability of the limonoate dehydrogenases from *Pseudomonas* and *Athrobacter*. *J. Agric. Food Chem.* **24**, 21.

Brodnitz, M. H., and Pascale, J. V. (1971). Thiopropanal-S-oxide a lachrymatory factor in onion. *J. Agric. Food Chem.* **19**, 269.

Brodnitz, M. H., Pascale, J. V., and Derslice, L. V. (1971). Flavor components of garlic extract. *J. Agric. Food Chem.* **19**, 273.

Brody, A. L., Bedrosian, K., and Ball, C. O. (1960). Low temperature handling of sterilized foods, V. Biochemical changes in storage. *Food Technol.* **14**, 552.

Brown, R. (1963). *In* "Cell Differentiation" (G. E. Fogg, ed.) Academic Press, New York pp. 1–17.

Brown, S. A. (1966). Lignins. *Annu. Rev. Plant Physiol.* **17**, 223.

Brown, S. A. (1969). Biochemistry of lignin formation. *BioScience* **19**, 115.

Brown, S. A., and Neish, A. C. (1955a). Shikimic acid as a precursor in lignin biosynthesis. *Nature (London)* **175**, 688.

Brown, S. A., and Neish, A. C. (1955b). Studies of lignin biosynthesis using isotopic carbon. IV. Formation from some aromatic isomers. *Can. J. Biochem. Physiol.* **33**, 948.

Brown, S. A., and Neish, A. C. (1956). Studies of lignin biosynthesis using isotopic carbon. V. Comparative studies on different plant species. *Can. J. Biochem. Physiol.* **34**, 769.

Brown, S. A., and Neish, A. C. (1959). Studies on lignin biosynthesis using isotopic carbon. VIII. Isolation of radioactive hydrogenolysis products of lignin. *J. Am. Chem. Soc.* **81**, 2419.

Bruemmer, J. H. (1969). Redox state of nicotinamide adenine dinucleotides in citrus fruit. *J. Agric. Food Chem.* **17**, 1312.

Buch, M. L., Satori, K. G., and Hills, C. H. (1961). The effect of bruising and aging on the texture and pectic constituents of canned red tart cherries. *Food Technol.* Chicago **15**, 526.

Buckle, K. A., and Edwards, R. A. (1970). Chlorophyll, color and pH changes in HTST processed green pea purée. *J. Fd. Technol.* **5**, 173.

Buttery, R. G., Seifer, R. M., Guadagni, D. G., and Ling, L. C. (1971). Characterization of additional volatile components of tomato. *J. Agric. Food Chem.* **19**, 524.

Byerrum, R. V., Flokstra, J. H., Dewey, L. J., and Ball, C. D. (1954). Incorporation of formate and the methyl group of methionine into methoxyl groups of lignin. *J. Biol. Chem.* **210**, 633.

Camm, E. L., and Towers, G. H. N. (1973a). Phenylalanine ammonia lyase. *Phytochemistry* **12**, 961.

Camm, E. L., and Towers, G. H. N. (1973b). Effect of aging on enzymes of phenylpropanoid metabolism in *Solanum tuberosum* disks. *Phytochemistry* **12**, 1575.

Campbell, H. (1937). Undesirable color changes in frozen peas stored at insufficiently low temperatures. *Food Res.* **2**, 55.

Carson, J. G., and Wong, F. F. (1961). The volatile flavor components of onions. *J. Agric. Food Chem.* **9**, 140.

Cavallito, C. J., Buck, J. S., and Suter, C. M. (1944). Allium the antibacterial principle of *Allium sativum*. II. Determination of chemical structure. *J. Am. Chem. Soc.* **66**, 1952.

Cavallito, C. J., Bailey, J. H., and Buck, J. S. (1945). The antibacterial principle of *Allium sativum*. III. Its precursor and essential oil of garlic. *J. Am. Chem. Soc.* **67**, 1032.

Challenger, F. (1953). The biological importance of organic compounds of sulfur. *Endeavour* **12**, 173.

Chanda, S. K., Hirst, E. L., and Perceval, E. G. V. (1951). The contributions of a pear cell wall xylan. *J. Chem. Soc.* p. 1240.

Chandler, B. V. (1971). A limonin-degrading enzyme in citrus albedo. *J. Sci. Food Agric.* 22, 634.

Chandler, B. V. (1977). One of the "101 most interesting problems in food science"—bitterness in orange juice, a case history. *Food Technol.* (Australia) **29**, 303.

Chandler, B. V., and Clegg, K. M. (1970). Pink discoloration in canned pears. I. Role of tin in pigment formation. *J. Sci. Food Agric.* **21**, 315.

Chandler, B. V., and Johnson, R. L. (1977). Cellulose acetate as a selective sorbent for limonin in orange juice. *J. Sci. Food Agric.* **28**, 875.

Chandler, B. V., Kefford, J. F., and Ziemelis, G. (1968). Removal of limonin from bitter orange juice. *J. Sci. Food Agric.* **19**, 83.

Chen, S. L., and Gutmanis, F. (1968). Auto-oxidation of extractable color pigments in Chili pepper with special reference to ethoxyquin treatment. *J. Food Sci.* **33**, 274.

Cho, Y.-P., and Chrispeels, M. J. (1976). Serine-O-galactosyl linkages in glycopeptides from carrot cell walls. *Phytochemistry* **15**, 165.

Chou, H., and Breene, N. M. (1972). Oxidative discoloration of β-carotene in low moisture foods. *J. Food. Sci.* **37**, 66.

Chrispeels, M. J. (1969). Synthesis and secretion of hydroxyproline containing macromolecules in carrots. I. Kinetic analysis. *Plant Physiol.* **44**, 1187.

Chrispeels, M. J., Sadava, D. and Cho, Y. P. (1974). Enhancement of extensin biosynthesis in aging disks of carrot storage tissue. *J. Ex. Bot.* **5**, 1157.

Clapp, R. C., Long, L. Jr., Dateo, G. P., Bissett, F. H., and Hasselström, J. (1959). The volatile isothiocyanates in fresh cabbage. *J. Am. Chem. Soc.* **81**, 6278.

Clore, W. J., Carter, G. H., and Drake, S. R. (1976). Pre and postharvest factors affecting textural quality of fresh asparagus. *J. Am. Soc. Hortic. Sci.* **101**, 576.

Clydesdale, F. M., and Francis, F. J. (1968). Chlorophyll changes in thermally processed spinach as influenced by enzyme conversion and pH adjustment. *Food Technol. (Chicago)* **22**, 793.

Clydesdale, F. M., Fleischmann, D. L., and Francis, F. J. (1970). Maintenance of color in processed green vegetables. *Food. Prod. Dev.* **4**, 127.

Coggins, C. W., Jr., and Knapp, J. C. F. (1969). Growth, development and softening of the Deglet Noor Date fruit. Date Grower Inst., Rep. **46**, 1.

Collins, J. L., and Wiley, R. C. (1963). Influence of added calcium salts on texture of thermal process apple slices. Univ. of Maryland Agr. Exp. Sta. Publ. NoA-130.

Creasy, L. L., and Zucker, M. (1974). Phenylalanine ammonia lyase and phenolic metabolism. *Recent Adv. Phytochem.* **8**. 1.

Crist, J. W., and Batjer, L. P. (1931). The stone cells of pear fruit especially the Kieffer pear. *Mich., Agric. Expt. Stn., Tech. Bull.* **113**, 1.

Croteau, R., and Loomis, D. W. (1975). Biosynthesis and metabolism of monoterpenes. *Int. Flav. Food Add.* 6, No. 5, 292.

Currier, H. B. (1957). Callose substances in plant cells. *Am. J. Bot.* **44**, 478.

Currier, H. B.. and Strugger, S. (1956). Aniline blue and fluorescence Microscopy of callose in bulb scales of *Allium cepa* L. *Protoplasma*, **45**, 552.

Currier, H. B., and Webster, D. H. (1964). Callose formation and subsequent disappearance: studies in ultrasound stimulation. *Plant Physiol.* **39**, 843.

Czaninski, Y., and Catesson, A. M. (1969). Localisation ultrastructurale d'activities peroxydasiques dans les tissus conducteurs végétaux au cours du cycle annuel. *J. Microsc. (Paris)* **7**, 875.

Czicki, V., and Kindl, H. (1977). Phenylalanine ammonia lyase and cinnamic acid hydroxylases as assembled consecutive enzymes on microsomal membranes of cucumber cotyledons: Co-operation and subcellular distribution. *Planta* **134**, 133.

Dalal, K. B. (1965). Ph.D. Thesis, Investigations into flavor chemistry with special reference to synthesis of volatiles in developing tomato fruit (*Lycopersicon esculentum*) under field and greenhouse growing conditions) Utah State University, Logan.

Dalal, K. B., Olson, L. E., Yu, M. H., and Salunkhe, D. K. (1967). Gas chromatography of the field-, glass- greenhouse-grown, and artificially ripened tomatoes (*Lycopersicon esculentum* Mill.). *Phytochemistry* **6**, 155.

Dalal, K. B.. Salunkhe, P. K., Olson. L. E., Do, J. Y., and Yu, M. H. (1968). Volatile components of developing tomato fruit grown under field and greenhouse conditions, *Plant Cell Physiol.* **9**, 389.

Daravingas, G., and Cain, R. F. (1968). Thermal degradation of black raspberry anthocyanin pigments in model systems. *J. Food Sci.* **33**, 138.

Dashek, W. V. (1970). Synthesis and transport of hydroxyproline-rich components in suspension cultures of sycamore-maple cells. *Plant Physiol.* **46**, 831.

Dashek, W. V., and Harwood, H. I. (1974). Proline, hydroxyproline and lily pollen tube elongation. *Ann. Bot. (London)* **38**, 947.

Dashek, W. V., and Rosen, W. G. (1966). Electron microscopical localization of chemical components in the growth zone by lily pollen tubes. *Protoplasma* **61**, 192.

Dateo, G. P., Clapp, R. C., MacKay, D. A. M., Hewitt, E. J., and Hasselstrom, T. (1957). Identification of the volatile sulfur components of cooked cabbage and the nature of the precursor in the fresh vegetable. *Food Res.* **22**, 440.

Datko, A. H., and Maclachan, G. A. (1968). Indoleacetic acid and synthesis of glucanases and pectic enzymes. *Plant Physiol.* **43**, 735.

Datta, S., and Nicholas, H. J. (1968). Incorporation of mevalonic acid 2-^{14}C into the triterpene limonin. *Phytochemistry* **7**, 955.

Davey, M. R., MacKenzie, I. A., Freeman, G. G.. and Short, K. C. (1974). Studies of some aspects of the growth, fine structure and flavor production of onion tissue grown *in vitro*. *Plant Sci. Lett.* **3**, 113.

Davis, R. B., Guyer, R. B., Daly, J. J., and Johnson, H. T. (1961). Control of rutin discoloration in canned asparagus. *Food Technol. Chicago* **15**, 212.

Decker, R. F. H., and Richards, G. N. (1976). Hemicellulases: Their occurrence, purification, properties, and mode of action. *Adv. Carbohydr. Chem.* **32**, 277.

DeJong, D. W. (1967). An investigation of the role of plant peroxidase in cell wall development by the histochemical method. *J. Histochem. Cytochem.* **15**, 335.

Dekazos, E. D. (1972). Callose formation by bruising and heating tomatoes and its presence in processed products. *J. Food Sci.* **37**, 562.

Dekazos, E. D., and Worley, J. F. (1967). Induction of callose formation by bruising and aging of red tart cherries. *J. Food Sci.* **32**, 287.

De La Mar, R. R., and Francis, F. J. (1969). Carotenoid degradation in bleached paprika. *J. Food Sci.* **34**, 282.

Delincee, H., and Radola, B. J. (1970). Some size and change properties of tomato pectinmethylesterase. *Biochim. Biophys. Acta* **214**, 178.

Della Monica, E. S., and McDowell, P. E. (1965). Comparison of beta-carotene content of dried carrots prepared by three dehydration processes. *Food Technol.* Chicago **19**, 141.

Demain, A. L., and Phaff, H. J. (1957). Recent advances in the enzymatic hydrolysis of pectic substances. *Wallerstein Lab. Commun.* **20**, 119.

De Swardt, G. H., and Maxie, E. C. (1967). Pectin methyl esterase in the ripening banana. *S. Afr. J. Agric. Sci.* **10**, 501.

Dickinson, D. B., and Hanson, J. B. (1965). Comparison of mitochondria from tomato fruits at various stages of ripeness. *Plant Physiol.* **40**, 161.

Dickinson, D. B., and McCollum, D. P. (1964). Cellulase in tomato fruits. *Nature (London)* **203**, 525.

Dietrich, W. C., Lindquist, F. E., Miers, J. C., Bohart, G. S., Neumann, H. J., and Tallourt, W. F. (1957). Objective tests to measure adverse changes in frozen vegetables. *Food Technol.* Chicago **11**, 109.

Dolev, A., Rohwedder, W. K., Mounts, T. L., and Dutton, H. J. (1967). A note on the mechanism of lipoxidase reaction and the origin of the oxygen incorporated into linoleate hydroperoxide. *Lipids* **2**, 442.

Drawert, F., Heiman, W., Emberger, R., and Tressl, R. (1966). Enzymatische Bildung von hexen-(2)-al(1), hexanal und deren Votstufen. *Justus Liebigs Ann. Chem.* **694**, 200.

Dreyer, D. L. (1966). Citrus bitter principles. VI. Ichangin. *J. Org. Chem.* **31**, 2279.

Dreyer, D. L. (1968). Limonoid bitter principles. *Fortschr. Chem. Org. Naturst.* **26**, 190.

Dunlap, W. J., and Wender, S. H. (1962). Identification studies on some minor flavonoid constituents of the grapefruit. *Anal. Biochem.* **4**, 110.

Dunlap, W. J., Hagen, P. E., and Wender, S. H. (1962). Preparation and properties of rhamnosidase and glucosidase fraction from a fungal flavonoid glucosidase preparation, naringinase C-100. *J. Food Sci.* **27**, 597.

Dunning, J. J. (1959). Factors influencing the formation of callose in plant cells. Ph.D. Thesis, University of California, Davis.

Edelman, J., and Hall, M. A. (1964). Effect of growth hormones on the development of invertase associated with cell walls. *Nature (London)* **201**, 296.

Eheart, M. S., and Odland, D. (1973). Use of ammonium compounds for chlorophyll retention in frozen green vegetables. *J. Food Sci.* **38**, 202.

Emerson, O. H. (1949). The bitter principle of navel oranges. *Food technol. (Chicago)* **3**, 248.

Eriksson, C. E. (1968). Alcohol: NAD oxidoreductase (EC.1.1.1.1.) from peas. *J. Food Sci.* **33**, 525.

Eskin, N. A. M., Henderson, H. M., and Townsend, R. J. (1971). "The Biochemistry of Foods." Academic Press, New York.

Ettlinger, M. G., and Lundeen, A. J. (1956). The structures of sinigrin and sinalbin; an enzymatic rearrangement. *J. Am. Chem. Soc.* **78**, 4172.

Evert, R. F., and Derr, W. F. (1964). Callose substance in sieve elements. *Am. J. Bot.* **51**, 552.

Finkle, B. J., and Masri, M. S. (1964). Methylation of polyhydroxyaromatic compounds by pampass grass O-methyltransferase. *Biochim. Biophys. Acta* **85**, 167.

Finkle, B. J., and Nelson, R. F. (1963). Enzyme reactions with phenolic compounds by *meta*-o-methyltransferase in plants. *Biochim. Biophys. Acta* **78**, 747.

Fisher, J. F. (1968). A procedure for obtaining radioactive naringin from grapefruit leaves fed L-phenylalanine ^{14}C. *Phytochemistry* **7**, 769.

Fisher, J. F., Nordby, H. E., and Kew, T. T. (1966). A thin-layer chromatographic-colorimetric procedure for determining naringin in grapefruit. *J. Food Sci.* **31**, 947.

Flavian, S., and Levi, A. (1970). A study of the natural disappearance of limonin-monolactone in the peel of Shamouti oranges. *J. Food Technol.* **5**, 127.

Fleming, H. P., Cobb, W. Y., Etchells, J. L., and Bell, T. A. (1968). The formation of carbonyl compounds in cucumbers. *J. Food Sci.* **33**, 572.

Foda, Y. H. (1957). Pectic changes during ripening as related to flesh firmness in tomato. Ph.D. Thesis, University of Illinois, Urbana.

Freeman, C. G., and Mossadeghi, N. (1970). Effect of sulfate nutrition on flavor components of onion (*Allium cepa*). *J. Sci. Food Agric.* **21**, 610.

Freeman, C. G., and Mossadeghi, N. (1973). Studies on the relationship between water regime and flavor strength between watercress (*Rorippa nasturium-aquaticum* (L) Hayek), cabbage (*Brassica oleracea capitata*) and onion (*Allium cepa*). *J. Hortic. Sci.* **48**, 365.

Freeman, C. G., and Whenham, R. J. (1975). A survey of volatile components of some *Allium* species in terms of S-alk (en) yl-L-cysteine sulfoxides present as flavor precursors. *J. Sci. Food Agric.* **26**, 1869.

Freeman, C. G., Whenham, R. J., MacKenzie, I. A., and Davey, M. R. (1974). Flavor components in tissue cultures of onion (*Allium cepa* L.). *Plant Sci. Lett.* **3**, 121.

Freeman, J. A., and Woodbridge, C. G. (1960). Effect of maturation, ripening and truss position on the free amino acid content in tomato fruits. *Proc. Am. Soc. Hortic. Sci.* **76**, 515.

Freudenberg, K. (1968). "Degradation of Lignin to Methoxybenzene carboxylic acids and other products. *In* "Constitution and Biosynthesis of Lignin. Part II" (K. Freudenberg and A. C. Neish, eds.), p. 86. Springer-Verlag, Berlin and New York.

Freudenberg, K., and Neish, A. C., eds. (1968). "Constitution and Biosynthesis of Lignin." Springer-Verlag, Berlin and New York.

Freudenberg, K., Reznik, H., Boesenberg, H., and Rasenack, D. (1952). Das an der Verholzung beteiligte Fermentsystem. *Chem. Ber.* **85**, 641.

Freudenberg, K., Seib, K., and Dall, K. (1959). Schwefehaltige Ligninpraparate und ihr vergleich mit modellsubstanzen. *Chem. Ber.* **92**, 807.

Frey-Wyssling, A. (1936). Der Aufbau der Pflanzichen Zellwande. *Protoplasma* **25**, 261.

Fridbourg, C. (1971). Growth and organogenesis in tissue cultures of *Allium cepa* var proliferum. *Physiol. Plant.* **25**, 436.

Furia, T. E., and Bellanca, N. (1975). Fenaroli's Handbook of Flavor Ingredients. 2nd Ed. Vol. 2, CRC Press, Inc.

Galliard, T., and Matthew, J. A. (1977). Lipoxygenase-mediated cleavage of fatty acids to carbonyl fragments in tomato fruits. *Phytochemistry* **16**, 339.

Galliard, T., Matthew, J. A., Wright, A. J., and Fishwick, M. J. (1977). The enzymic breakdown of lipids to volatile and non-volatile carbonyl fragments in disrupted tomato fruits. *J. Sci. Food Agric.* **28**, 863.

Gardiner, M., and Chrispeels, M. J. (1975). Involvement of the Golgi apparatus in the synthesis and secretion of hydroxyproline-rich cell wall glycoproteins. *Plant Physiol.* **55**, 536.

Gardner, H. W., and Clagett, C. O. (1965). Linoleate hydroperoxide decomposition (enzymatic) by extracts of alfalfa seedling extracts. *Plant Physiol.* **40**, Suppl., XVII.

Gauch, J. G., and Eaton, G. M. (1942). Effect of saline substrate on hourly levels of carbohydrates and inorganic constituents of barley plants. *Plant Physiol.* **17**, 347.

George-Nascimento, C., and Cori, O. (1971). Terpene biosynthesis from geranly and neryl pyrophosphates by enzymes from orange flavedo. *Phytochemistry* **10**, 1083.

Gieseker, L. F. (1949). Art of preserving and maintaining color of green vegetables. U.S. Patent 2, 437, 747.

Gilliver, P. J., and Nursten, H. E. (1976). The source of the acyl moiety in the biosynthesis of volatile banana esters. *J. Sci. Fd. Agric.* **27**, 152.

Gini, B., and Koch, R. B. (1961). Study of a lipohydroperoxide breakdown factor in soy extracts. *J. Food Sci.* **26**, 359.

Gold, H. J., and Weckel, K. G. (1959). Degradation of chlorophyll to pheophytin during sterilization of canned peas by heat. *Food Technol. Chicago* **13**, 280.

Goldstein, J. L., and Wick, E. L. (1969). Lipids in ripening banana fruit. *J. Food Sci.* **34**, 482.

Goldstein, L., Lifshitz, A., and Sokolovsky, M. (1971). Water soluble derivatives of naringinase. *Int. J. Biochem.* **2**, 448.

Graham, R. C., Jr., and Karnovsky, M. J. (1966). The early stages of absorption of injected horseradish peroxidase in the proximal tubule of mouse kidney: Ultra-structural cytochemistry by a new technique. *J. Histochem. Cytochem.* **14**, 291.

Granroth, B. (1970). Biosynthesis and decomposition of cysteine derivatives in onion and other Allium species. *Ann. Acad. Sci. Fenn. Ser. A2.* **154**, 40–46.

Gregor, H. D. (1976). Studies on phenylalanine ammonia lyase from castor bean endosperm. 1. Subcellular localization and induction of the enzyme. *Z. Pflanzenphysiol.* **77**, 454.

Gremli, H., and Wild, J. (1974). Enzymatic flavor regeneration in processed foods. *Proc. Int. Congr. Food Sci. Technol. 4th,* Vol. I, p. 158–161.

Griffiths, F. P., and Lime, B. J. (1959). Debittering of grapefruit products with naringinase. *Food Technol. (Chicago)* **13**, 430.

Grommeck, R., and Markakis, P. (1964). Effect of peroxidase on anthocyanin pigments. *J. Food Sci.* **29**, 53.

Gross, G. G., Stoeckigt, J., Mansell, R. L., and Zenk, M. H. (1973). Three novel enzymes involved in the reduction of ferulic acid to coniferyl alcohol in higher plants; ferulate; CoA ligase, feruloyl-CoA reductase and coniferyl alcohol oxidoreductase. *FEBS Lett.* 31, 283.

Guadagni, D. G., Buttery, R. G., and Venstrom, D. N. (1972). Contribution of some volatile tomato components and their mixtures to fresh tomato aroma. *J. Sci. Food Agric.* 23, 1445.

Guadagni, D. G., Maier, V. P., and Turnbaugh, J. G. (1973). Effect of some citrus juice constituents on taste thresholds for limonin and naringin bitterness. *J. Sci. Food Agric.* **24**, 1277.

Gupte, S. M., and Francis, F. J. (1964). Effect of pH adjustment and high-temperature short-time processing on color and pigment retention in spinach puree. *Food Technol. (Chicago)* **18**, 141.

Gupte, S. M., El-Bisi, H. M., and Francis, F. J. (1964). Kinetics of thermal degradation of chlorophyll in spinach puree. *J. Food Sci.* **29**, 379.

Guymon, J. F. (1965). Mutant strains of *Saccharomyces cerevisiae* applied to studies of higher alcohol formation during fermentation. *Dev. Ind. Microbiol.* 7, 88.

Hagen, R. E., Dunlap, W. J., and Wender, S. H. (1966). Seasonal variation of naringin and certain other flavonone glycosides in juice sacs of Texas Ruby Red Grapefruit. *J. Food Sci.* 31, 542.

Hahlbrock, K., and Grisebach, H. (1970). Formation of coenzyme A esters of cinnamic acids with an enzyme preparation from cell suspension cultures of parsley. *FEBS Lett.* **11**, 62.

Hall, C. B. (1963). Cellulase in tomato fruits. *Nature (London)* **200**, 1010.

Hall, C. B. (1964). Cellulase activity in tomato fruits according to portion and maturity. *Bot. Gaz. (Chicago)* **125**, 156.

Hall, C. B., and Dennison, R. A. (1960). The relationship of firmness and pectinesterase activity of tomato fruits. *Proc. Am. Soc. Hortic. Sci.* **75**, 629.

Hall, C. B., and Mullins, J. T. (1965). Cellulase activity in sterile tomato fruits. *Nature (London)* **206**, 638.

Hamson, A. R. (1952). Measuring firmness of tomatoes in a breeding program. *Proc. Am. Soc. Hortic. Sci.* **60**, 425.

Hanson, K. R., Zucker, M., and Sondheimer, E. (1967). *In* "Phenolic Compounds and Metabolic Regulation" (B. J. Furkle and V. C. Runeckles, eds.), p. 68. Appleton-Century-Crofts, New York.

Harkin, P. M., and Obst, J. R. (1973). Lignification in trees: Indication of exclusive peroxidase participation. *Science* **180**, 296.

Harper, K. A. (1968). Structural changes of flavylium salts. IV. Polarographic and spectrometric examination of pelargonidin chloride. *Aust. J. Chem.* **21**, 221.

Hasegawa, S. (1976). Metabolism of limonoids, Limonin D-ring lactone hydrolase activity in *Pseudomonas. J. Agric. Food Chem.* **24**, 124.

Hasegawa, S., and Hoagland, J. E. (1977). Biosynthesis of limonoids in citrus. *Phytochemistry* **16**, 469.

Hasegawa, S., and Maier, V. P. (1970). Biosynthesis of *trans*-cinnamate from phenyl pyruvate and L-glutamate by cell-free extracts of grapefruit. *Phytochemistry* **9**, 2483.

Hasegawa, S., and Maier, V. P. (1972). Cinnamate hydroxylation and the enzymes leading from phenyl pyruvate to *p*-coumarate synthesis in grapefruit tissues. *Phytochemistry* **11**, 1365.

Hasegawa, S., and Smolensky, D. C. (1971). Cellulase in dates and its role in softening. *J. Food Sci.* **36**, 966.

Hasegawa, S., Maier, V. P., Kaszycki, H. P., and Crawford, J. K. (1969). Polygalacturonase content of dates and its relation to maturity and softness. *J. Food Sci.* **34**, 527.

Hasegawa, S., Bennett, R. D., Maier, V. P., and Douglas, King, A., Jr. (1972). Limonoate dehydrogenase from *Arthrobacter globiformis. J. Agric. Food Chem.* **20**, 1031.

Hasegawa, S., Maier, V. P., and Bennett, R. D. (1974a). Detection of limonoate dehydrogenase activity in albedo tissues of *Citrus sinenses. Phytochemistry* **13**, 103.

Hasegawa, S., Maier, V. P., and Douglas, King, A., Jr. (1974b). Isolation of new limonoate dehydrogenase from *Pseudomonas. J. Agric. Food Chem.* **22**, 523.

Hasegawa, S., Maier, V. P., Border, S. N., and Bennett, R. D. (1974c). Metabolism of Limonoids. Isolation and characterization of deoxylimonin hydrolase from *Pseudomonas. J. Agric. Food Chem.* **22**, 1093.

Hatanaka, A., Kajiwari, T., and Harada, T. (1975). Biosynthetic pathway of cucumber alcohol: *Trans*-2, *cis*-6-nonadienol via *cis*-3, *cis*-6-nonadienal. *Phytochemistry* **14**, 2589.

Heinz, D. E., Creveling, R. K., and Jennings, W. G. (1965). Direct determination of aroma compounds as an index of pear maturity. *J. Food Sci.* **30**, 641.

Henrici, M. (1952). Comparative study of the content of starch and sugars of *Tribulus terristris* Lucerne, some *Graiminea* and *Pentzia* under different meterological, edaphical, and physiological conditions, Paper 2, Carbohydrate nutrition. *Onderstepoort J. Vet. Sci.* **25**, 45 [Cited by Freeman, G. G., and Mossadghi, N. (1973). *J. Hortic. Sci.* **48**, 365.]

Hepler, P. K., and Fosket, D. E. (1971). The role of microtubules in vessel member differentiation in *Coleus. Protoplasma* **72**, 213.

Hepler, P. K., Rice, R. M., and Terranova, W. A. (1972). Cytochemical localization of peroxidase activity in wound vessel members of *Coleus*. *Can. J. Bot.* **50**, 977.

Herner, R. C. (1973). Fiber determination. *In* Progress report of asparagus research. Mich. State Univ. Agr. Sta. Res. Rpt. 217, p. 11–12.

Hess, D. (1965a). Comparison of the methylating potencies of genotypes with differently methylated anthocyanins in a cell-free system. *Z. Pflanzenphysiol.* **53**, 1; cited in *Chem. Abstr.* **65**, 17382a (1966).

Hess, D. (1965b). Methylation of 4-hydroxyferulic acid to sinapic acid in a cell-free system. *Z. Pflanzenphysiol.* **53**, 460; cited in *Chem. Abstr.* **65**, 17378h (1966).

Hess, K., Mahl, H., and Gutter, E. (1957). Electron microscopic representation of long periodic intervals in cellulose fibers and comparison with the periods of other kinds of fibers. *Kolloid-Z.* **155**, 1.

Hewitt, E. J., MacKay, D. A. M., Konigsbacher, K. S., and Hasselstrom, T. (1957). Flavor propagation through enzyme action. Chemistry of Natural Flavors. A symposium sponsored by the National Academy of Sciences—National Research Council, for the Quartermaster Food and Container Institute for the Armed Forces, and Pioneering Research Division, Quartermaster Research and Engineering Center. Ser. 9, pp. 86–100.

Higby, R. H. (1938). The bitter constituents of Navel and Valencia Oranges. *J. Am. Chem. Soc.* **60**, 3013.

Higuchi, T. (1957). Biochemical studies of lignin formation. I. *Physiol. Plant.* **10**, 356.

Higuchi, T. (1958). Further studies on phenol oxidase related to the lignin biosynthesis. *J. Biochem. (Tokyo)* **45**, 515.

Higuchi, T., and Brown, S. A. (1963). Studies on lignin biosynthesis using isotopic carbon. XIII. The phenoylpropanoid system in lignification. *Can. J. Biochem. Physiol.* **41**, 621.

Higuchi, T., Kawamura, I., Morimoto, I., and Kimura, N. (1954). Formation of lignin in the bamboo shoot. III. Lignin formation from precursors by a section of bamboo shoot tissue. *J. Jpn. For. Soc.* **36**, 253; cited in *Chem. Abstr.* **50**, 3566i (1956).

Higuchi, T., Kawamura, I., Yamazaki, N., and Morimoto, I. (1955). Formation of lignin in bamboo shoot. V. Inhibitory effect of ascorbic acid-glutathione system on the oxidative condensation of coniferyl alcohol by peroxidase. *J. Jpn. For. Soc.* **37**, 502; cited in *Chem. Abstr.* **50**, 8809b (1956).

Hijwegen, T. (1963). Lignification, a possible mechanism of active resistance against pathogens. *Neth. J. Plant Pathol.* **69**, 314.

Hill, A. C., and Rhodes, M. J. C. (1975). The properties of cinnamic acid 4-hydroxylase of aged swede root disks. *Phytochemistry* **14**, 2387.

Hills, C. H., Whittenberger, R. T., Robertson, W. F., and Case, W. H. (1953). Studies on the processing of red cherries. II. Some effects of bruising on yield and quality of canned Montmorency cherries. *Food Technol. (Chicago)* **7**, 32.

Hobson, G. E. (1963). Pectinesterase in normal and abnormal tomato plants. *Biochem. J.* **86**, 358.

Hobson, G. E. (1964). Polygalacturonase in normal and abnormal tomato fruit. *Biochem. J.* **92**, 324.

Hobson, G. E. (1965). The firmness of tomato fruit in relation to polygalacturonase activity. *J. Hortic. Sci.* **40**, 66.

Hobson, G. E. (1967). The effects of alleles at the "never ripe" locus on the ripening of tomato fruit. *Phytochemistry* **6**, 1337.

Hobson, G. E. (1968). Cellulase activity during the maturation and ripening of tomato fruit. *J. Food Sci.* 33, 588.

Hobson, G. E. (1962). Pectic enzymes in the ripening of normal and abnormal fruits. *Food Sci. Technol., Proc. Int. Congr., 1st,* Vol. 1, p. 461.

Holden, M. (1965). Chlorophyll bleaching by legume seeds. *J. Sci. Food Agric.* **16**, 312.

Holt, C., and Sellen, D. B. (1975). The statistics of chained sectioning experiments on cellulose fibers. *J. Polym. Sci.* **13**, 1.

Horowitz, R. M. (1964). Relationships between the taste and structure of some phenolic compounds. *In* "Biochemistry of Phenolic Compounds" (J. B. Harborne, ed.), Chapter 14, p. 545. Academic Press, New York.

Horowitz, R. M., and Gentili, B. (1969). Taste and structure in phenolic glycosides. *J. Agric. Food Chem.* **17**, 697.

Hsu, A. C., Hasegawa, S., Maier, V. P., and Bennett, R. D. (1973). 17-dehydrolimonoate A-ring lactone in citrus fruits. *Phytochemistry* **12**, 563.

Huang, H. T. (1955). Decolorization of anthocyanins by fungal enzymes. *J. Agric. Food Chem.* 3, 141.

Huang, H. T. (1956). The kinetics of the decolorization of anthocyanins by fungal enzyme. *J. Am. Chem. Soc.* **78**, 2390.

Hultin, H. O. (1959). Flavor and aroma constituents of banana and banana purée. Ph.D. Thesis, Massachusetts Institute of Technology, Boston.

Hultin, H. O., and Levine, A. S. (1965). Pectinmethylesterase in the ripening banana. *J. Food Sci.* **30**, 917.

Hultin, H. O., and Proctor, B. E. (1961). Changes in some volatile constituents of the banana during ripening, storage and processing. *Food Technol. (Chicago)* **15**, 440.

Hultin, H. O., and Proctor, B. E. (1962). Banana aroma precursors. *Food Technol. (Chicago)* **16**, 108.

Hunter, W. J., and Elkan, G. H. (1974). Endopolygalacturonase from tomato fruit. *Phytochemistry* **13**, 2725.

Imamura, M., and Shimizu, S. (1974). Metabolism of chlorophyll in higher plants. IV. Relationship of fatty acid oxidation and chlorophyll bleaching in plant extracts. *Plant Cell Physiol.* **15**, 187.

Isherwood, F. A. (1963). Lignin, Recent Advances in Food Science. Vol. 3, pp. 300–309. (J. M. Leitch, and D. N. Rhodes, eds.). Butterworths, London.

Issenberg, P., and Wick, E. L. (1963). Volatile components of bananas. *J. Agric. Food Chem.* **11**, 2.

Jadhav, S., Singh, B., and Salunkhe, D. K. (1972). Metabolism of unsaturated fatty acids in tomato fruit: Linoleic and linolenic acid as precursors of hexanal. *Plant Cell Physiol.* **13**, 449.

Jensen, K. A., Corti, J., and Kjaer, A. (1953). Isothiocyanates. II. Volatile isothiocyanates in seeds and roots of various *Brassicae*. *Acta Chem. Scand.* 7, 1267.

Jeremyn, M. A., and Isherwood, F. A. (1956). Changes in the cell wall of pears during ripening. *Biochem. J.* **64**, 123.

Joffe, F. M., Stier, E. F., Bongolan, D. C., Epstein, A. I., and Ball, C. O. (1961). Low temperature handling of sterilized foods. IV. Color and flavor of canned vegetables. *Food Technol. (Chicago)* **15**, 507.

Johnson, J. H., Gould, W. A., Badenhop, A. F., and Johnson, R. M., Jr. (1968). Quantitative comparison of isoamylol, pentanol and 3-hexenol-1 in tomato juice. Varietal and harvest differences and processing effects. *J. Agric. Food Chem.* **16**, 255.

Jones, I. D., White, R. C., and Gibbs, E. (1961). The formation of pheophorbides during brine preservations of cucumber. *Food Technol. (Chicago)* **15**, 172.

Jones, I. D., White, R. C., and Gibbs, E. (1963). Influence of bleaching or brining treatments on the formation of chlorophyllides, pheophytins and pheophorbides in green plant tissue. *J. Food Sci.* **28**, 437.

Joslyn, M. A. (1952). Protopectin. *Adv. Food Res.* **11**, 1.

Joslyn, M. A., and Mackinney, G. (1938). The rate of conversion of chlorophyll to pheophytin. *J. Am. Chem. Soc.* **60**, 1132.

Joyce, A. E. (1954). Some polyenes from Brassica rutabaga. *Nature (London)* **173**, 311.

Jurd, L., and Asen, S. (1966). The formation of metal and co-pigment complexes of cyanidin-3-glucoside. *Phytochemistry* **5**, 1263.

Kanemitsu, T., and Matsuura, F. (1961). Studies on the carotenoids of salmon. II. On the fading of the red color in canned salmon. *Bull. Jpn. Soc. Sci. Fish.* **27**, 453.

Kapp, P. P. (1966). Some effects of variety, maturity, and storage on fatty acids in fruit pericarp of *Lycopersicon esculentum* Mill. *Diss. Abstr.* **27**, 77B.

Karr, A. L., Jr. (1972). Isolation of an enzyme system which will catalyze the glycosylation of extensin. *Plant Physiol.* **50**, 275.

Karymann, F. (1967). The effect of the rate and duration of growth on the amount of fiber in green asparagus spears. *Arch. Gartenbay* **15**, 89.

Kazeniac, S. J., and Hall, R. M. (1970). Flavor chemistry of tomato volatiles. *J. Food Sci.* **35**, 519.

Keegstra, K., Talmadge, K. W., Bauer, W. D., and Albersheim, P. (1973). The structure of plant cell walls. III. A model of the walls of suspension-cultured cells based on the interconnections of the macromolecular components. *Plant Physiol.* **51**, 188.

Kefford, J. F. (1959). The chemical constituents of citrus fruits. *Adv. Food Res.* **9**, 285.

Kefford, J. F., and Chandler, B. V. (1970). The chemical constituents of citrus fruits. *Adv. Food Res.* Suppl. 2, p. 150.

Kertesz, Z. I. (1938). Pectic enzymes. II. Pectic enzymes of tomatoes. *Food Res.* 3, 481.

Kertesz, Z. I. (1939). The effect of calcium on plant tissues. *Canner* **85**, 26.

Kertesz, Z. I. (1951). "The Pectic Substances." Wiley (Interscience), New York.

Kertesz, Z. I., Tolman, T. G., Loconti, J. D., and Ruyle, C. H. (1940). The use of calcium in the commercial canning of whole tomatoes. N.Y. State Agr. Exp. Sta. Tech. Bull. 252.

Kessler, G., Feingold, D. S., and Hassid, W. Z. (1960). Utilization of exogenous sugars for biosynthesis of carbohydrates in germinating pollen. *Plant Physiol.* 35, 505.

Khan, M. M. R. (1952). A study of "lipoxidase" in the flesh of British Columbia Herring. *J. Fish Res. Bd.* Canada, 9,393.

Kies, M. W., Haining, J. L., Pistorius, E., Schroeder, D. H., and Axelrod, B. (1969). On the question of the identity of soybean "lipoxidase" and carotene oxidase. *Biochem. Biophys. Res. Commun.* 36, 312.

King, K. W., and Vessal, M. J. (1969). Enzymes of the cellulase complex. *Adv. Chem. Ser.* **95**, 7.

Klaui, H. (1971). Carotinoide in Lebensmitteln. Presented at the Symposium on "Functional Properties in Foods" sponsored by the Swiss Institute of Food Technology and the Swiss Federal Institute of Technology, Switzerland.

Kleber, C. (1912). The occurrence of amylacetate in bananas. *Am. Perfum.* 7, 235.

Knee, M. (1974). Change in structural polysaccharides of apple ripening during storage. *Colloq. Int.* No. 238, Factors et Regulation de la Maturation des Fruits, Centre de la Recherche Scientifique, Paris.

Knox, R. B. (1972). New methods for overcoming cross-incompatability in plants. *Incompatability Newsl.* **1**, 15.

Knox, R. B. (1973). Pollen wall proteins; pollen-stigma interactions in ragweed and *Cosmos* (Compositae). *J. Cell Sci.* **12**, 241.

Koblitz, H. (1961). Einige Bermerkungen zuon histochemuschen nachweis van an der Verholzung bereiligten Ferrenten. *Ber. Dtsch. Bot. Ges.* **74**, 274.

Koch, R. B. (1962). Dehydrated foods and model systems. *In* "Lipids and Their Oxidation" (H. W. Schultz, ed.), p. 230. Avi Publ. Co., Westport, Connecticut.

Kokoul, J., and Conn, E. G. (1961). Metabolism of aromatic compounds in higher plants. IV. Purification and properties of the phenylalanine deaminase of *Hordeum vulgare*. *J. Biol. Chem.* **236**, 2962.

Konigsbacher, V. S., Hewitt, E. J., and Evans, R. L. (1959). Applications of flavor enzymes to processed foods. I. Panel studies. *Food Technol.* Chicago **13**, 128.

Kramer, A., and Twigg, B. A. (1956). Quality control for the canner and freezer. *Canner Freezer* **122**, 16.

Kramer, A., Haut, I. C., Scott, L. E., and Ide, L. E. (1949). Objective methods for measuring quality factors of raw, canned and frozen asparagus. *Proc. Am. Soc. Hortic. Sci.* **76**, 382.

Kratky, O. (1935). The micellar structure and the deformation processes of fibrous substances. *Kolloid-Z.* **70**, 14.

Kuroda, H., Shimada, M., and Higuchi, T. (1975). Purification and properties of O-methyltransferase involved in the biosynthesis of gymnosperm lignin. *Phytochemistry* **14**, 1749.

LaBelle, R. L., and Moyer, J. C. (1960). Factors affecting the drained weight and firmness of red tart cherries. *Food Technol. (Chicago)* **14**, 347.

LaJollo, F., Tannenbaum, S. R., and Labuza, T. P. (1971). Reaction at limited water concentration. 2. Chlorophyll degradation. *J. Food Sci.* **36**, 850.

Lamport, D. T. A. (1962). Hydroxyproline of primary cell walls. *Fed. Am. Soc. Exp. Biol.* **21**, 398.

Lamport, D. T. A. (1965). The protein component of primary cell walls. *Adv. Bot. Res.* **2**, 151.

Lamport, D. T. A. (1967). Hydroxyproline-O-glycosidic linkage of the plant cell wall glycoprotein extensin. *Nature (London)* **216**, 1322.

Lamport, D. T. A. (1970). Cell wall metabolism. *Annu. Rev. Plant Physiol.* **21**, 235.

Lamport, D. T. A., Katona, L., and Roerig, R. (1973). Galactosylserine in extensin. *Biochem. J.* **133**, 125.

Lamport, D. T. A., and Miller, D. H. (1971). Hydroxyproline-arabinosides in the plant kingdom. *Plant Physiol.* **48**, 454.

Lamport, D. T. A., and Northcote, D. H. (1960). Hydroxyproline in primary cell walls of higher plants. *Nature (London)* **188**, 665.

Laties, G. G. (1964). The relation of glucose absorption to respiration in potato slices. *Plant Physiol.* **39**, 391.

Lee, M., and Macmillan, J. D. (1968). Mode of action of pectic enzymes. 1. Purification and certain properties of tomato pectinesterase. *Biochemistry* **7**, 4005.

Leeper, G. F. (1973). Pectin ultrastructure. I. Pectate elementary fibrils. *J. Texture Stud.* **4**, 248.

Lerch, G. (1960). Untersuchungen über Wurzelkallose. *Bot. Stud.* **11**.

Li, L. H., Flora, R. M., and King, K. W. (1963). Purification of the betaglucosidases from

Aspergillus niger and initial observations on the C_1 of *Trichoderma koningii*. *J. Ferment. Technol.* 41, 98.

Lipetz, J., and Garro, A. J. (1965). Ionic effects of lignification and peroxidase in tissue cultures. *J. Cell Biol.* 25, 109.

Loconti, J. D., and Kertesz, Z. I. (1941). Identification of calcium-pectate as the tissue-firming compound formed by treatment of tomatoes with calcium chloride. *Food Res.* 6, 499.

Loef, H. W., and Thung, S. B. (1965). Uberden Einflub von Chlorophyllase auf die Farbe von Spincet Wahrend und nacht der verwertung. *Z. Lebensm.-Unters. -Forsch.* 126, 401.

Luckner, M. (1972). "Secondary Metabolism in Plants and Animals," p. 261. Chapman & Hall, London.

Lueck, R. H. (1970). Black discoloration in canned asparagus: Interrelations of iron, tin, oxygen and rutin. *J. Agric. Food Chem.* 18, 607.

Luh, B. S., Leonard, S. J., and Phaff, H. J. (1956). Hydrolysis of pectic materials and oligouronides by tomato polygalacturonase. *Food. Res.* 21, 448.

Luh, B. S., Leonard, S. J., and Patel, D. S. (1960). Pink discoloration in canned Bon Chrétien Pears. *Food Technol. (Chicago)* 14, 53.

Luh, B. S., Leonard, S. J., Simone, M., and Villarreal, F. (1964). Asceptic canning of foods. VII. Strained lima beans. *Food Technol. (Chicago)* 18, 363.

Lukton, A., Chichester, C. O., and MacKinney, G. (1956). The breakdown of strawberry anthocyanin pigment. *Food Technol.* Chicago 10, 427.

McCalla, D. R., and Neish, A. C. (1959). Metabolism of phenylpropanoid compounds in *Salvia*. II. Biosynthesis of phenolic cinnamic acids. *Can. J. Biochem. Physiol.* 37, 537.

McCarthy, A. I., Palmer, J. K., Shaw, C. P., and Anderson, E. A. (1963). Correlation of gas chromatographic data with flavor profiles of fresh banana fruit. *J. Food Sci.* 28, 379.

McClendon, J. H. (1964). Evidence for the pectic nature of the middle lamella of potato tuber cell walls based on chromatographic macerating enzymes. *Am. J. Bot.* 51, 628.

McColloch, R. J. (1950). Preliminary studies on debittering Navel orange products. *Calif. Citrogr.* 35, 290.

McColloch, R. J., Keller, G. T., and Beavens, E. A. (1952). Factors influencing quality of tomato products. 1. Surfaced localized pectic enzymes inactivated by blanching. *Food Technol. (Chicago)* 6, 197.

McCready, B. M., and McComb, E. A. (1954). Pectic constituents in ripe and unripe fruit. *Food Res.* 19, 530.

MacKay, D. A. M., and Hewitt, E. J. (1959). Application of flavor enzymes to processed foods. II. Comparison of the effect of flavor enzymes from mustard and cabbage upon dehydrated cabbage. *J. Sci. Food Agric.* 19, 273.

Mackinney, G., and Weast, C. A. (1940). Color changes in green vegetables. *Ind. Eng. Chem.* 22, 392.

Mackinney, G., Lukton, A., and Greenbaum, L. (1958). Carotenoid stability in stored dehydrated carrots. *Food Technol. (Chicago)* 12, 164.

Maclachan, G. A., and Perrault, J. (1964). Cellulose from pea epicotyls. *Nature (London)* 204, 81.

MacLeod, A. J. (1970). The chemistry of vegetable flavor. *Flavour Ind.* 1, 665.

MacLeod, A. J., and MacLeod, G. (1968). Volatiles of cooked cabbage. *J. Sci. Food Agric.* 19, 273.

MacLeod, A. J., and MacLeod, G. (1970a). The flavor volatiles of dehydrated cabbage. *J. Food Sci.* **35**, 739.

MacLeod, A. J., and MacLeod, G. (1970b). Effects of variations in cooking methods on the flavor volatiles of cabbage. *J. Food Sci.* **35**, 744.

MacLeod, W. D. (1966). Nootkatone, grapefruit flavor and the citrus industry. *Calif. Citrogr.* **51**, 120.

McNairn, R. B., and Currier, H. B. (1968). Translocation blockage of sieve plate callose. *Planta* **82**, 369.

Maier, V. P. (1969). Compositional studies in citrus significance in processing, identification and flavor. *Proc. Int. Citrus Symp., 1st, 1969,* p. 235.

Maier, V. P., and Beverly, G. D. (1968). Limonin monolactone the non-bitter precursor responsible for delayed bitterness of certain orange juices. *J. Food Sci.* **33**, 488.

Maier, V. P., and Dreyer, D. L. (1965). Citrus bitter principles. IV. Occurrence of limonin in grapefruit juice. *J. Food Sci.* **30**, 874.

Maier, V. P., and Hasegawa, S. (1970). L-phenylalanine ammonia-lyase activity and naringenin glycoside accumulation in developing grapefruit. *Phytochemistry* **9**, 139.

Maier, V. P., and Margileth, D. A. (1969). Limonoic acid A-ring lactone, a new limonin derivative in citrus. *Phytochemistry* **8**, 243.

Maier, V. P. and Metzler, D. M. (1967a). Grapefruit phenolics. I. Identification of dihydrokaemferol and its occurrence with naringin and kaemferol. *Phytochemistry* **6**, 763.

Maier, V. P., and Metzler, D. M. (1967b). Grapefruit phenolics. II. Principal aglycones of endocarp and peel and possible biosynthetic relationship. *Phytochemistry* **6**, 1127.

Maier, V. P., Hasegawa, S., and Hera, E. (1969). Limonin D-ring-lactone hydrolase, a new enzyme from citrus seeds. *Phytochemistry* **8**, 405.

Maier, V. P., Brewster, L. C., and Hsu, A. C. (1971). Development of methods for producing non-bitter Navel orange juice. *Calif. Citrogr.* **56**, 351.

Maier, V. P., Brewster, L. C., and Hsu, A. C. (1973). Ethylene-accelerated limonoid metabolism in citrus fruits: A process for reducing juice bitterness. *J. Agric. Food Chem.* **21**, 490.

Major, R. T., and Thomas, M. (1972). Formation of 2-hexenal from linolenic acid by macerated Ginkgo leaves. *Phytochemistry* **11**, 611.

Major, R. T., Marchini, P., and Boulton, T. (1963). Observations of the production of α-hexenal by leaves from certain plants. *J. Biol. Chem.* **238**, 1813.

Major, R. T., Marchini, P., and Sproston, T. (1968). Isolation from *Ginkgo biloba* L. of an inhibitor of fungus growth. *J. Biol. Chem.* **235**, 3298.

Maloney, J. F., Labuza, T. P., Wallace, D. H., and Karel, M. (1966). Autoxidation of methyl linoleate in freeze-dried model systems. I. Effect of water on the autocatalyzed oxidation. *J. Food Sci.* **31**, 878.

Manley, R. St. John. (1964). Fine structure of native cellulose microfibrils. *Nature (London)* **204**, 1155.

Mann, J., Roldan-Gonzalez, L., and Wellard, H. J. (1960). Crystalline modifications of cellulose. Part IV. Determination of X-ray intensity data. *J. Polym. Sci.* **42**, 165.

Mansell, R. L., Stoeckigt, J., and Zenk, M. H. (1972). Reduction of ferulic acid to coniferyl alcohol in a cell free system from a higher plant. *Z. Pflanzenphysiol.* **68**, 286.

Mapson, L. W., and Moustafa, E. M. (1955). The oxidation of glutathione by a lipoxidase enzyme from pea seeds. *Biochem. J.* **60**, 71.

Markakis, P., Livingstone, C. E., and Fellers, C. R. (1957). Quantitative aspects of strawberry pigment degradation. *Food Res.* **22**, 117.

Martinez, F., and Labuza, T. P. (1968). Rate of deterioration of freeze-dried salmon as a function of relative humidity. *J. Food Sci.* 33, 241.

Marx-Figini, M., and Schultz, G. V. (1966). Uber die kinetik und den mechanismus der biosynthese der cellulose in den hoheren pflanzen (nach versuchen an den samenhaaren der baumwolle). *Biochim. Biophys. Acta* **112**, 81.

Mason, H. S., and Cronyn, M. (1955). On the role of polyphenoloxidase in lignin biosynthesis. *J. Am. Chem. Soc.* 77, 491.

Mathews, A. G. (1969). Some characteristics of the additional anthocyanidins formed during the conversion of leucoanthocyanidins into anthocyanidins. *Phytochemistry* **8**, 627.

Matikkala, E. J., and Virtanen, A. I. (1965). γ-glutamylpeptidase (Glutaminase) in germinating seeds of chives. *Acta Chem. Scand.* **19**, 1258.

Matikkala, E. J., and Virtanen, A. I. (1967). On the quantitative determination of the amino acids and γ-glutamylpeptides of onions. *Acta Chem. Scand.* **21**, 2891.

Meigh, D. F., Pratt, H. K., and Cole, C. (1966). Production of carbonyl compounds by tomato fruit tissue in the presence of aliphatic alcohols. *Nature (London)* **211**, 419.

Melera, A., Schaffner, K., Arigoni, D., and Jeger, O. (1957). Zur Konstitution des Limonins. I. Uber den Verlauf der alkalischen Hydrolys von Limonin and Limonol. *Helv. Chim. Acta* **40**, 1420.

Meschter, E. E. (1953). Effects of carbohydrates and other factors on strawberry products. *J. Agric. Food Chem.* **1**, 574.

Meyer, K. H., and Misch, L. (1937). The constitution of the crystallite part of cellulose. VI. The positions of the atoms in the new space model of cellulose. *Helv. Chim. Acta* **20**, 232.

Minimakawa, T., and Uritani, I. (1965). Phenylalanine ammonia-lyase in sliced sweet potato roots. *J. Biochem. (Tokyo)* **57**, 678.

Mitchell, H. L., and Hauge, S. M. (1976). Enzymic nature of the carotene-destroying system of alfalfa. *J. Biol. Chem.* **163**, 7.

Mohammadzadeh-Khayat, A., and Luh, B. S. (1968). Calcium and oxalate ions effect on the texture of canned apricots. *J. Food Sci.* 33, 493.

Moisio, T., Spare, C. G., and Virtanen, A. I. (1962). Mass spectral studies of the chemical nature of the lachrymatory factor formed enzymically from S(1-propenyl)-cysteine sulfoxide isolated from onion. *Suom. Kemistil. B* **35**, 29.

Monro, J. A., Penny, D., and Bailey, R. W. L. (1976). The organization and growth of primary cell walls of lupin hypocotyl. *Phytochemistry* **15**, 1193.

Monties, B. (1974). Sur la présence de phenylalaninéaminonialase dans les chloroplastes des cellules follares. *C.R. Hebd. Seances Acad. Sci.* **278**, 1465.

Mori, Y., Hirayama, M., and Tokunaga, Y. (1966). Formation of the organic acids and the fading of chlorophyll during the storage of green vegetables. *Eiyo To Shokuryo* b119, 191.

Muggli, R., Elias, H. G., and Mühlethaler, K. (1969). Zum Feinbau der Elementarfibrillen der cellulose. *Makromol. Chem.* **121**, 290.

Mühlethaler, K. (1969). Die Feinstruktor der Zellulosemikrofibrillen. *Schweiz. Z. Forstwes. Beih.* **30**, 55.

Mühlethaler, K. (1969). Fine structure of natural polysaccharide systems. *J. Polym. Sci. C* **28**, 305.

Murray, K. E., Palmer, J. K., Whitfield, F. G., Kennett, B. H., and Stanley, G. (1968). The volatile alcohols of ripe bananas. *J. Food Sci.* 33, 632.

Myers, M. J., Issenberg, P. I., and Wick, E. L. (1969). Vapor analysis of the production by banana of certain fruit volatile constituents. *J. Food Sci.* **34**, 504.

Myers, M. J., Issenberg, P. I., and Wick, E. L. (1970). L-leucine as a precursor of iso-amyl alcohol and iso-amyl acetate. Volatile aroma constituents of banana fruit discs. *Phytochemistry* **9**, 1693.

Nagai, I. (1921). A genetico-physiological study on the formation of anthocyanins and brown pigments in plants. *J. Coll. Agric., Hokkaido Imp. Univ.* **8**, 2.

Nagel, C. W., and Patterson, M. E. (1967). Pectic enzymes and development of the pear (*Pyrus communis*). *J. Food Sci.* **32**, 294.

Nair, P. M., and Vining, L. C. (1965). Cinnamic acid hydroxylase in spinach. *Phytochemistry* **4**, 161.

Nakamura, W. (1967). Studies on the biosynthesis of lignin. I. Disproof against the catalytic activity of laccase in the oxidation of coniferyl alcohol. *J. Biochem. (Tokyo)* **62**, 54.

Narayan, C. V., Stout, B. A., Norris, E. R., and Cargill, B. F. (1969). Physical properties of asparagus spears. Tech. Seminar on Asparagus Harvest Mechanization. *Amer. Soc. Eng.*

Neish, A. C. (1961). Formation of *m*- and *p*-coumaric acids by enzymatic deamination of the corresponding isomers of tyrosine. *Phytochemistry* **1**, 1.

Neish, A. C. (1968). Monomeric Intermediates in the Biosynthesis of Lignin. *In* "Constitution and Biosynthesis of Lignin. Part I" (K. Freudenberg and A. C. Neish, eds.), p. 7. Springer-Verlag, Berlin and New York.

Nelson, P. E., and Hoff, J. E. (1969). Tomato volatiles: Effect of variety, processing and storage time. *J. Food Sci.* **33**, 290.

Nieduszynski, I., and Preston, R. D. (1970). Crystallite size in natural cellulose. *Nature (London)* **225**, 273.

Northcote, D. H. (1963). The biology and chemistry of the cell walls of higher plants, algae and fungi. *Int. Rev. Cytol.* **14**, 223.

Northcote, D. H. (1969). The synthesis and metabolic control of polysaccharides and lignin during differentiations of plant cells. *Essays Biochem.* **5**, 89.

Northcote, D. H. (1972). Chemistry of the plant cell wall. *Annu. Rev. Plant Physiol.* **23**, 13.

Nortje, B. K. (1964). Factors relating to the discoloration of canned Bon Chrétien pears. *S. Afr. J. Agric. Sci.* **9**, 681.

Nozu, Y. (1968). Studies on the biosynthesis of lignin. III. Dehydrogenative polymerization of coniferyl alcohol by peroxidase. *J. Biochem. (Tokyo)* **62**, 519.

Nye, W., and Spoehr, H. A. (1943). The isolation of hexenal from leaves. *Arch. Biochem. Biophys.* **2**, 23.

Oaks, E. M., Hartmann, H., and Dimick, K. P. (1964). Analysis of sulfur compounds with electron capture hydrogen flame dual channel chromatography. *Anal. Chem.* **36**, 1560.

Omura, H., Ishizaki, K., Osajima, Y., and Yamafuji, K. (1963). Naringinase activity of some enzyme preparation. *Kyushi Daigaku Nogakubu Gakugei Zasshi* **20**, 169; *Chem. Abstr.* **61**, 4639e (1966).

Ono, T., and Nagayama, F. (1957a). Studies on the lipoxidase in fish tissues. I. Oxidation of vitamin A by the extract of some fish liver. *Bull. Jpn. Soc. Sci. Fish.* **23**, 206.

Ono, T., and Nagayama, F. (1957b). Studies on the lipoxidase in fish tissues. II. Lipoxidase action of liver and dark muscle. *Bull. Jpn. Soc. Sci. Fish.* **23**, 211.

Ono, T., and Nagayama, F. (1959). Studies on the lipoxidase in fish tissues. III. Linoleate oxidation by liver extract. *Bull. Jpn. Soc. Sci. Fish.* **24,** 853.

Orlowski, M., and Meister, A. (1965). Isolation of γ-glutamyl transpeptidase from hog kidney. *J. Biol. Chem.* **240,** 1338.

Paillard, N. (1968). Analyse de l'arôme de pommes de la variété Calville blanc par chromatographic sur colonne capillaire. *Fruits* **23,** 283.

Palmer, J. K. (1971). The banana. *In* "Biochemistry of Fruits and Their Products" (A. C. Hulme, ed.), Vol. 2, p. 65. Academic Press, New York.

Palmer, J. K. (1973). Separation of components of aroma concentrates on the basis of functional group and aroma quality. *J. Agric. Food Chem.* **21,** 923.

Palmer, J. K., and Roberts, J. B. (1967). Inhibition of banana polyphenoloxidase by 2-mercaptobenzothiazole. *Science* **157,** 200.

Palmer, J. K., Shaw, T. M., and Ballantyne, M. (1947). An X-ray diffraction investigation of pectinic and pectic acids. *J. Phys. Chem.* **51,** 710.

Parish, R. W. (1972). The intracellular location of phenol oxidases, peroxidase and phosphatases in the leaves of Spinach, Beet (*Beta vulgaris* L. Subspecies vulgaris). *Eur. J. Biochem.* **31,** 446.

Patel, D. S., and Phaff, H. J. (1960a). Studies on the purification of tomato polygalacturonase. *Food Res.* **25,** 37.

Patel, D. S., and Phaff, H. J. (1960b). Properties of purified tomato polygalacturonase. *Food Res.* **25,** 47.

Patschke, L., Bartz, W., and Grisebach, H. (1966). Stereospezifischer Einbau von (−) 5.7.4-Trihydroxyflavonon in Flavonoide and Isoflavone. *Z. Naturforsch., Tail B* **21,** 137.

Peng, C. Y., and Markakis, P. (1963). Effect of phenolase on anthocyanins. *Nature (London)* **199,** 597.

Pharr, D. M., and Dickinson, D. B. (1973). Partial characterization of Cx cellulase and cellobiase from ripening tomato fruits. *Plant Physiol.* **51,** 577.

Pifferi, P. G., and Cultrera, R. (1974). Enzymatic degradation of anthocyanins: The role of sweet cherry polyphenol oxidase. *J. Food Sci.* **39,** 786.

Pinsky, A., Grossman, S., and Trop, M. (1971). Lipoxygenase content and antioxidant activity of some fruits and vegetables. *J. Food Sci.* **36,** 571.

Pithawala, H. R., Sovur, G. R., and Sreenivas, A. (1968). Characterization of tomato pectinesterase. *Arch. Biochem.* 17, 235.

Polgar, A., and Zechmeister, L. (1942). Isomerization of β-carotene. Isolation of a stereoisomer with increased adsorptive affinity. *J. Chem. Soc.* **64,** 1856.

Pope, D. G. (1977). Relationships between hydroxyproline-containing proteins secreted into the cell wall and mediums by suspension-cultured *Acer polydoplatanus* cells. *Plant Physiol.* **59,** 894.

Potts, J. R. M., Weklych, R., and Conn, E. E. (1974). The 4-hydroxylation of cinnamic acid by Sorghum microsomes and the requirement for cytochrome P-450. *J. Biol. Chem.* **249,** 5019.

Potty, V. H. (1969). Occurrence and properties of enzymes associated with mevalonic acid synthesis in the orange. *J. Food Sci.* **34,** 231.

Potty, V. H., and Bruemmer, J. H. (1970a). Mevalonate activating enzymes in the orange. *Phytochemistry* **9,** 99.

Potty, V. H., and Bruemmer, J. H. (1970b). Oxidation of geraniol by an enzyme system from orange. *Phytochemistry* **9,** 1003.

Potty, V. H., and Bruemmer, J. H. (1970c). Formation of isoprenoid pyrophosphates from mevalonate by orange enzymes. *Phytochemistry* **9**, 1229.

Potty, V. H., and Bruemmer, J. H. (1970d). Limonene reductase system in the orange. *Phytochemistry* **9**, 2319.

Poux, N. (1969). Localisation d'activités enzymatiques dans les cellules du meristère gradiculaire de *Cucumis sativus* L. II. Activité peroxydasique. *J. Microsc. (Paris)* **8**, 855.

Pratt, D. E., Balkom, C. M., Powers, J. J., and Mills, L. W. (1954). Fruit color stability, interaction of ascorbic acid, riboflavin and anthocyanin pigments. *J. Agric. Food Chem.* **2**, 367.

Preil, W. (1970). Observing the growth of pollen tubes in pistil and ovarian tissue by means of fluorescence microscopy. *Zeiss Inf. Bull.* **75**, 25.

Pressey, R., and Avants, J. K. (1971). Effects of substrate size on the activity of tomato polygalacturonases. *J. Food Sci.* **36**, 486.

Pressus, R., and Avants, J. K. (1972). Multiple forms of pectinesterase in tomatoes. *Phytochemistry* **11**, 3139.

Pressey, R., and Avants, J. K. (1973). Separation and characterization of endopolygalacturonase and exopolygalacturonase from peaches. *Plant Physiol.* **52**, 252.

Pressey, R., and Avants, J. K. (1975a). Modes of action of carrot and peach exopolygalacturonases. *Phytochemistry* **14**, 957.

Pressey, R., and Avants, J. K. (1975b). Cucumber polygalacturonase. *J. Food Sci.* **40**, 937.

Pressey, R., and Avants, J. K. (1976). Pear polygalacturonases. *Phytochemistry* **15**, 1349.

Pressey, R., and Avants, J. K. (1977). Occurrence and properties of polygalacturonases in *Avena* and other plants. *Plant Physiol.* **60**, 458.

Pressey, R., Hinton, D. M., and Avants, J. K. (1971). Development of polygalacturonase activity and solubilization of pectin in peaches during ripening. *J. Food Sci.* **36**, 1070.

Preston, R. D. (1971). Negative staining and cellulose microfibril size. *J. Microsc. (Oxford)* **93**, 7.

Preston, R. D., and Cronshaw, J. (1958). Constitution of the fibrillar and nonfibrillar components of the walls of *Valania ventriosa*. *Nature (London)* **181**, 288.

Pryke, J. A., and ap Rees, T. (1976). Activity of the pentose phosphate pathway during lignification. *Planta* **132**, 279.

Pryke, J. A., and ap Rees, T. (1977). The pentose phosphate pathway as a source of NADPH for lignin synthesis. *Phytochemistry* **16**, 557.

Purcell, A. E. (1962). Carotenoids of Goldrush sweet potato flakes. *Food Technol. (Chicago)* **16**, 99.

Purcell, A. E., Walters, W. M., Jr., and Thomkins, W. T. (1971). Relationship of vegetable color to physical state of the carotenes. *J. Agric. Food Chem.* **17**, 41.

Pyne, A. H., and Wick, E. L. (1965). Volatile components of tomatoes. *J. Food Sci.* **30**, 192.

Ranadive, A. S., and Haard, N. F. (1972a). Chemical nature of stone cells from pears. *32nd Ann. Meet. Inst. Food Technol.*

Ranadive, A. S., and Haard, N. F. (1972b). Peroxidase localization and lignin formation in developing pear fruit. *J. Food Sci.* **27**, 381.

Ranby, B. (1969). Recent progress on the structure and morphology of cellulose. *Adv. Chem. Ser.* **99**, 139.

Rasmussen, G. K. (1973). Changes in cellulase and pectinase activities in fruit tissues and separation zones of citrus treated with cycloheximide. *Plant Physiol.* **51**, 626.

Rasmussen, G. K. (1974). Cellulase activity in separation zones of citrus fruit treated with abscisic acid under normal and hypobaric atmospheres. *J. Amer. Soc. Hort. Sci.* **99**, 229.

Rasmussen, G. K. (1975). Cellulase activity, endogenous absicisic acid and ethylene in four citrus cultivars during maturation. *Plant Physiol.* **56**, 765.

Rasmussen, G. K., and Jones, J. W. (1971). Cellulase activity and ethylene concentration of citrus fruit sprayed with abscission-inducing chemicals. *HortScience* **6**, 402.

Rawlins, T. G., and Takahashi, W. N. (1952). "Technics of Plant Histochemistry and Virology." Natl. Press, Millbrae, California.

Reese, D. A. (1969). Structure, conformation and mechanism in the formation of polysaccharide gels and networks. *Adv. Carbohyd. Chem.* **24**, 267.

Reese, E. T., Siu, R. G. H., and Levinson, H. S. (1950). Biological degradation of soluble cellulase derivatives and its relationship to the mechanism of cellulase hydrolysis. *J. Bacteriol.* **59**, 485.

Reeve, R. M. (1970). Relationships of histological structure to texture of fresh and processed fruits and vegetables. *J. Texture Stud.* **1**, 247.

Reymond, D., and Phaff, H. J. (1965). Purification and certain properties of avocado polygalacturonases. *J. Food Sci.* **30**, 266.

Reynolds, J. D., and Dashek, W. V. (1976). Cytochemical analysis of callose localization in *Lilium longiflorum* pollen tubes. *Ann. Bot. (London)* **40**, 409.

Rhee, K. S., and Watts, B. W. (1966). Effects of antioxidants on lipoxidase activity in model systems and pea (*Pisum sativum*) slurries. *J. Food Sci.* **31**, 669.

Rhodes, M. J. C., and Wooltorton, L. S. C. (1971). The effect of ethylene on the respiration and on the activity of phenylalanine ammonia lyase in swede and parsnip root tissue. *Phytochemistry* **10**, 1989.

Rhodes, M. J. C., and Wooltorton, L. S. C. (1973). Formation of CoA esters of cinnamic acid derivatives by extracts of *Brassica napo-brassica* root tissues. *Phytochemistry* **12**, 2381.

Rhodes, M. J. C., and Wooltorton, L. S. C. (1974). Reduction of the CoA thioesters of p-coumaric and ferulic acids by extracts of aged *Brassica napo-brassica* root tissue. *Phytochemistry* **13**, 107.

Rhodes, M. J. C., and Wooltorton, L. S. C. (1975). Enzymes involved in the reduction of ferulic acid to coniferyl alcohol during the aging of disks of swede root tissue. *Phytochemistry* **14**, 1235.

Rhodes, M. J. C., Hill, A. C. R., and Wooltorton, L. S. C. (1976). Activity of enzymes involved in lignin biosynthesis in swede root disks. *Phystochemistry* **15**, 707.

Ride, J. P. (1975). Lignification in wounded wheat leaves in response to fungi and its possible role in resistance. *Physiol. Plant Pathol.* **5**, 183.

Roelofsen, P. A. (1953). Polygalacturonase activity in yeast, Neurospora and tomato extract. *Biochim. Biophys. Acta* **10**, 410.

Roelofsen, P. A., and Kreger, D. R. (1951). The submicroscopic structure of pectin in collenchyma cell-walls. *Exp. Bot.* **2**, 232.

Roggen, H. P., J. R. (1972). Scanning electron microscopical investigations revealed the presence of a wax later on the stigma puppliae of *Brassica. Incompatability Newsl.* **1**, 25.

Romani, R. J., and Ku, L. (1968). Direct gas chromatographic analysis of volatiles produced by ripening fruit. *J. Food Sci.* **31**, 558.

Rothenbach, F., and Eberlein, L. (1905). Ubër das vorkommen von Estem in den Fruchten der Bananen. *Deut. Essigind* **4**, 81.

Rosen, W. G., and Gawlik, S. R. (1966). Fine structure of lily pollen tube tips following various fixation and staining procedures. *Protoplasma* **61**, 181.

Rouse, A. H., and Barmore, C. R. (1974). Changes in pectic substances during ripening of avocadoes. *HortScience* **9**, 36.

Ruis, H., and Kindl, H. (1970). Distribution of ammonia-lyases in organelles of castor bean endosperm. *Hoppe-Seyler's Z. Physiol. Chem.* **351**, 1425.

Ruis, H., and Kindl, H. (1971). Formation of α, β unsaturated carboxylic acids from amino acids in plant peroxisomers. *Phytochemistry* **10**, 2627.

Russel, D. W. (1971). The metabolism of aromatic compounds in higher plants. X. Properties of the cinnamic acid 4-hydroxylase of pea seedlings and some aspects of its metabolic and developmental control. *J. Biol. Chem.* **246**, 3870.

Russel, D. W., and Conn, E. E. (1967). The cinnamic acid 4-hydroxylase of pea seedlings. *Arch. Biochem. Biophys.* **122**, 256.

Ryugo, K. (1969). Seasonal trends in titratable acidity, tannins and polyphenolic compounds and cell wall constituents in oriental pear fruit. (*P. serotina*). *J. Agric. Food Chem.* **17**, 43.

Sadava, D., and Chrispeels, M. J. (1971). Hydroxyproline biosynthesis in plant cells. Peptidyl proline hydroxylase from carrot disks. *Biochim. Biophys. Acta* **227**, 278.

Saghir, A. R., and Mann, L. K. (1965). The lachrymatory factor in American wild alliums. *Suom. Kemistil. B* **38**, 78.

Saghir, A. R., Mann, L. K., Bernhard, R. A., and Jacobson, J. V. (1964). Determination of aliphatic mono and disulfides in *Allium* by gas chromatography and their distribution in the common food species. *Proc. Am. Soc. Hortic. Sci.* **84**, 386.

Sakamura, S., Watanabe, S. and Obata, V. J. (1965). Anthocyanase and anthocyanins occurring in eggplant. 3. Oxidative decoloration of the anthocyanin by polyphenol oxidase. *Agr. Biol. Chem.* **29**, 181.

Sakamura, S., Shibusa, S., and Obata, Y. J. (1966). Separation of a polyphenol oxidase for anthocyanin degradation in eggplant. *J. Food Sci.* **31**, 317.

Salmon, J. (1947). Différentiation des tubes eribles chez les angiospermes. Recherches cytologique. *Rev. Cytol. Cytophysiol. Veg.* **9**, 55.

Salwin, H. (1959). Defining minimum moisture contents for dehydrated foods. *Food Technol. (Chicago)* **13**, 594.

Sato, M. (1966). Metabolism of phenolic substances by the chloroplasts of p-coumaric acid to coffeic acid. *Phytochemistry* **5**, 385.

Schanderl, S. H., Chichester, C. O., and Marsh, B. L. (1962). Degradation of chlorophyll and several derivatives in acid solution. *J. Org. Chem.* **27**, 3865.

Schanderl, S. H., Marsh, G. L., and Chichester, C. O. (1965a). Color reversion in processed vegetables. I. Studies on regreened pea purée. *J. Food Sci.* **30**, 312.

Schanderl, S. H., Marsh, G. L., and Chichester, C. O. (1965b). Color reversion in processed vegetables. II. Model system studies. *J. Food Sci.* **30**, 317.

Schopfer, P. (1971). Phenylalanine ammonia lyase (PAL, EC 4.3.1.5) of *Sinapis alba* (mustard) seedling, an elecrophoretically homogenous enzyme. *Planta* **99**, 339.

Schormuller, J., and Grosch, W. (1964). Aromatus in foods. 1. Identification of carbonyl compounds in tomatoes. *Z. Lebensm.-Unters.-Forsch.* **118**, 385.

Schormuller, J., and Grosch, W. (1965). Untersuchungen über Aromastoffe von Lebensmitteln. 2. Über das Vorkommen weiterer Carbonylverbindungen in der Tomate. *Z. Lebensm.-Unters.-Forsch.* **126**, 38.

Schwimmer, S. (1963). Alteration of the flavor of processed vegetables by enzyme preparations. *J. Food Sci.* **28,** 760.

Schwimmer, S. (1968). Enzymatic conversion of *trans* (+)-S-1-propenyl-L-cysteine sulfoxide to the bitter and odor bearing components of onion. *Phytochemistry* **7,** 401.

Schwimmer, S. (1969). Characterisation of S-propyl-L-cysteine sulfoxide as the principal endogenous substrate of L-cysteine sulfoxide lyase of onion. *Arch. Biochem. Biophys.* **130,** 312.

Schwimmer, S. (1971a). Enzymatic conversion of γ-L-glutamyl cysteine peptides to pyruvic acid, a coupled reaction for enhancement of onion flavor. *J. Agric. Food Chem.* **19,** 980.

Schwimmer, S. (1971b). Flavor enhancement of allium products. U.S. Patent 3,725,085.

Schwimmer, S., and Austin, S. J. (1971a). Gamma glutamyl transpeptidase of sprouted onion. *J. Food Sci.* **36,** 807.

Schwimmer, S., and Austin, S. J. (1971b). Enhancement of pyruvic acid release and flavor in dehydrated *Allium* powders by gamma glutamyl transpeptidases. *J. Food Sci.* **36,** 1081.

Schwimmer, S., and Friedman, M. (1972). Genesis of volatile sulfur-containing food flavors. *Flavour Ind.* **3,** 137.

Schwimmer, S., and Guadagni, D. G. (1962). Relation between olfactory threshold concentration and pyruvic acid content of onion juice. *J. Food Sci.* **27,** 94.

Schwimmer, S., and Guadagni, D. G. (1968). Kinetics of the enzymatic development of pyruvic acid and odor in frozen onions treated with cysteine C-S lyase. *J. Food Sci.* **33,** 193.

Schwimmer, S., and Weston, W. J. (1961). Enzymatic development of pyruvic acid in onion as a measure of pungency. *J. Agric. Food Chem.* **9,** 301.

Segerlind, L. J., and Herner, R. C. (1972). On the fiber content problem of processed asparagus. *Am. Soc. Agric. Eng.* **72,** 882.

Selby, K. (1969). The purification and properties of the C_1 component of the cellulase complex. *Adv. Chem. Ser.* **95,** 34.

Semmler, F. W. (1892). The essential oil of onion. (*Allium cepa* L.). *Arch. Pharm. (Weinheim, Ger.)* **230,** 443.

Sha, B. J., Salunkhe, D. K., and Olson, L. E. (1969). Effects of ripening process on chemistry of tomato volatiles. *J. Am. Soc. Hortic. Sci.* **94,** 171.

Shafizadeh, F., and McGinnis, G. D. (1971). Morphology and biogenesis of cellulose and plant cell walls. *Adv. Carbohyd. Chem.* **26,** 297.

Shankaranarayana, J. L., Raghavan, B., Abraham, K. O., and Natarayan, C. P. (1974). Volatile sulfur compounds in food flavors. *Crit. Rev. Food Technol.* **4,** 395.

Shimada, M., Yamazaki, R., and Higuchi, T. (1970a). Metabolism of p-coumaric acid during lignification of a bamboo. *Phytochemistry* **9,** 1.

Shimada, M., Ohasi, H., and Higuchi, T. (1970b). O-methyl transferases involved in the biosynthesis of lignins. *Phytochemistry* **9,** 2463.

Shimada, M., Fushiki, H., and Higuchi, T. (1972). O-methyltransferase activity from Japanese black pine. *Phytochemistry* **11,** 2657.

Shimizu, C., and Fukuhara, T. (1957a). Influence of the liver enzyme on the decrease of vitamin A in the fish liver during storage. *Bull. Jpn. Soc. Sci. Fish.* **23,** 115.

Shimizu, C., and Fukuhara, T. (1957b). The specificity of lipoxidase-like enzyme in mackerel liver. *Bull. Jpn. Soc. Sci. Fish.* **23,** 174.

Shimokoriyama, H. (1957). Interconversion of chalcones and flavonones of a phloroglucinol-type structure. *J. Am. Chem. Soc.* **79**, 4199.

Shimomura, T., and Dijkstra, J. (1975). The occurrence of callose during the process of local lesion formation. *Neth. J. Plant Pathol.* **81**, 107.

Siegel, A., Markakis, P., and Bedford, C. L. (1971). Stabilization of anthocyanins in frozen tart cherries by blanching. *J. Food Sci.* **36**, 962.

Singleton, V. L., Gortner, W. A., and Yang, H. Y. (1961). Carotenoid pigments of pineapple fruit. I. Acid-catalyzed isomerization of the pigments. *J. Food. Sci.* **26**, 49.

Sistrunk, W. A., and Cash, R. (1968). Stabilizing the color of frozen strawberries. *Arkansas Farm Res.* **17**, 2.

Smith, H. R., and Kramer, A. (1947). The fiber content of canned asparagus. *The Canner* **104**, 14.

Smith, M. M., and McCully, M. E. (1977). Mild temperature "Stress" and callose synthesis. *Planta* **136**, 65.

Sobotka, F. E., and Stelzig, A. A. (1974). An apparent cellulase complex in tomato (*Lycopersicon esculentum L.*) fruit. *Plant Physiol.* **53**, 759.

Sobotka, F. E., and Watada, A. G. (1971). Cellulase in high pigment and crimson tomato fruit. *Proc. Am. Soc. Hortic. Sci.* **96**, 705.

Sondheimer, E. (1953). On the relation between spectral changes and pH of the anthocyanin pelargonidin-3-monoglucoside. *J. Am. Chem. Soc.* **75**, 1507.

Sondheimer, E., and Kertesz, Z. I. (1952). The kinetics of the oxidation of strawberry anthocyanin by hydrogen peroxide. *Food Res.* **17**, 288.

Sondheimer, E., and Kertesz, Z. I. (1953). Participation of ascorbic acid in the destruction of anthocyanin in strawberry juice and model systems. *Food Res.* **18**, 475.

Sosa-Coronel, J., Vest, G., and Hemer, R. C. (1976). Distribution of fiber content in asparagus cultivars. *HortScience* **11**, 149.

Spalding, D. H., and Reeder, W. F. (1972). Quality of 'Booth 8' and 'Lula' avocadoes stored in controlled atmospheres. *Proc. Fla. State Hortic. Soc.* **85**, 337.

Spare, C. G., and Virtanen, A. I. (1963). On the lachrymatory factor in onion (*Allium cepa*) vapors and its precursors. *Acta Chem. Scand.* **17**, 641.

Spencer, M. S., and Stanley, W. L. (1954). Flavor and odor components in the tomato. *J. Agric. Food Chem.* **2**, 1113.

Stafford, H. A. (1974). The metabolism of aromatic compounds. *Annu. Rev. Plant Physiol.* **25**, 459.

Stafford, H. A. (1975). Transformations of a large aggegate of hydroxycinnamate hydroxylase to lower molecular weight forms by sulfhydryl agents in green leaves of *Sorghum. Plant Physiol.* **55**, 145.

Stafford, H. A. (1976). Characteristics of a 4-hydroxycinnamate hydroxylase purified from *Sorghum* leaves. *Plant Physiol.* **57**, 320.

Stafford, H. A., and Baldy, R. (1970). A monophenol oxidase activity in extracts of *Sorghum. Plant Physiol.* **45**, 215.

Stafford, H. A., and Dresler, S. (1972). 4-Hydroxycinnamic acid hydroxylase and polyphenoloxidase activities in *Sorghum vulgare. Plant Physiol.* **52**, 453.

Stanley, W. L. (1948). *In* "Flavor Research and Food Acceptance" (L. B. Sjoström, ed.), p. 344. Van Nostrand-Reinhold, Princeton, New Jersey.

Starr, M. S., and Francis, F. J. (1968). Oxygen and ascorbic acid effect on the relative stability of four anthocyanin pigments in cranberry juice. *Food Technol. (Chicago)* **22**, 91.

Sterling, C. (1961). Physical state of cellulase during the ripening of peach. *J. Food Sci.* **26**, 95.

Stevens, M. A. (1970). Inheritance and flavor contribution of 2-isobutylthiazole, methyl salicylate and eugenol in tomatoes. *J. Am. Soc. Hortic. Sci.* **95**, 9.

Steward, F. C., Hulme, A. C., Frieberg, S. R., Hegarty, M. P., Pollard, J. K., Rabson, R., and Barr, R. A. (1960). Physiological investigations on the banana plant. II. Factors which affect the nitrogen compounds of the fruit. *Ann. Bot. (London)* **24**, 117.

Stewart, C. R. (1967). The effects of water and oxygen content on changes in amino acid content of excised leaves and studies on proline metabolism in wilted and unwilted leaves. *Diss. Abstr. B* **28**, 25; cited in *Hortic. Abstr.* **38**, 4718 (1968).

Stoll, A., and Seebeck, E. (1948). The isolation of sinigrin as the true crystallised mother substance of horse-radish oil. *Helv. Chim. Acta* **31**, 1432.

Stone, E. J., Hall, R. M., and Kazeniac, S. J. (1971). Precursors of tomato volatiles. *Pap., 162nd Natl. Meet. Am. Chem. Soc.* Abstract No. 31 Division of Agricultural and Food Chemistry American Chemical Society.

Stone, E. J., Hall, R. M., and Kazeniac, S. J. (1975). Formation of aldehydes and alcohols in tomato fruit from U-^{14}C-labelled linolenic and linoleic acids. *J. Food Sci.* **40**, 1138.

Strain, H. H. (1941). Unsaturated fat oxidase: Specificity, occurrence and induced oxidations. *J. Am. Chem. Soc.* **63**, 3542.

Sugano, N., Iwata, R., and Nishi, A. (1975). Formation of phenolic acid in carrot cells in suspension cultures. *Phytochemistry* **14**, 1205.

Sumner, R. J. (1943). Lipoid oxidase studies. Method for the determination of lipoxidase activity. *Ind. Eng. Chem. Anal. Ed.* **15**, 14.

Sutter, A., and Grisebach, H. (1969). Biosynthesis of flavonoids. XXXIV. Occurrence of the NIH shift in flavonoid biosynthesis. *Phytochemistry* **8**, 101.

Suzuki, M., Motohiro, T., and Tanikawa, E. (1969). Color assessment of canned red salmon *(Oncorhynchus nerka). FAO Tech. Conf. Fish Inspection Qual. Control, 1969* Topic III, No. FE.FIC/69/0/58

Sweeney, J. P., and Martin, M. G. (1961). Stability of chlorophyll in vegetables as affected by pH. *Food Technol. (Chicago)* **15**, 263.

Swirski, M. A., Allouf, R., Guimard, A., and Cheftel, H. (1969). A water soluble, stable green pigment originating during processing of canned brussel sprouts picked before the first Autumn frosts. *J. Agric. Food Chem.* **17**, 799.

Synge, R. L. M., and Wood, J. C. (1956). (+)-(S-methyl-L-cysteine S-oxide) in cabbage. *Biochem. J.* **64**, 252.

Szczesniak, A. S., and Kleyn, D. H. (1963). Consumer awareness of texture and other food attributes. *Food Technol. (Chicago)* **17**, 74.

Talmadge, K. W., Keegstra, K., Bauer, W. D., and Albersheim, P. (1973). The structure of plant cell walls. I. The macromolecular components of the walls of suspension-cultured sycamore cells with a detailed analysis of pectic polysaccharides. *Plant Physiol.* **51**, 158.

Tan, C. T., and Francis, F. J. (1962). Effect of processing temperature on pigments and color of spinach. *J. Food Sci.* **27**, 232.

Tanimoto, E., and Masuda, Y. (1968). Effect of auxin on cell wall degrading enzymes. *Physiol. Plant.* **21**, 820.

Tavakoli, M., and Wiley, R. C. (1968). Relation of trimethylsilyl derivatives of fruit tissue polysaccharides to apple texture. *Proc. Am. Soc. Hortic. Sci.* **92**, 780.

Thomas, D. W., Smyth, C. V., and Labbee, M. P. (1958). Enzymic hydrolysis of naringin, the bitter principle of grapefruit. *Food Res.* **23**, 591.

Thung, S. B. (1963). Browning of heated spinach. *Annu. Rep., Inst. Res. Storage Process. Hortic. Produce* p. 43.

Ting, S. W. (1958). Enzymic hydrolysis of naringin in grapefruit. *J. Agric. Food. Chem.* **6**, 546.

Tinsley, I. J., and Bockian, A. H. (1960). Some effects of sugar on the breakdown of pelargonidin-3-glucoside in model systems at 90°C. *Food Res.* **25**, 161.

Tokoroyama, T., and Matsuura, T. (1962). Formation of methyl tetrahydroanhydroepimero-obacunolate from limonin. *Nippon Kagaku Zasshi* **83**, 630; *Chem. Abstr.* **49**, 534.

Tressl, R., and Drawert, F. (1973). Biogenesis of banana volatiles. *J. Agric. Food. Chem.* **21**, 560.

Tressl, R., and Jennings, W. G. (1972). Production of volatile compounds in the ripening banana. *J. Agric. Food Chem.* **20**, 190.

Tressl, R., Drawert, F., and Heimann, W. (1970). Uber die Biogenese von Aromastoffen bei Pflanzen und Fruchten. V. Anreicherung, Trennung und Identifizierrung von Banaenaromastoffen. *Z. Lebensm.-Unters.-Forsch.* **142**, 249.

Tsukuda, N., and Amano, K. (1966). Studies on the discoloration of red fishes. I. Content of carotenoid pigments in eighteen species of red fishes. *Bull. Jpn. Soc. Sci. Fish.* **32**, 334.

Tsukuda, N., and Amano, K. (1967). Studies on the discoloration of red fishes. IV. Discoloration of astaxanthin, tunaxanthin and β-carotene by the tissue homogenates of fishes. *Bull. Jpn. Soc. Sci. Fish.* **33**, 962.

Tsukuda, N., and Amano, K. (1968). Studies on the discoloration of red fishes. V. Enzyme involved in the discoloration of carotenoid pigments in fish skin tissues. *Bull. Jpn. Soc. Sci. Fish.* **34**, 633.

Tupy, J. (1959). Callose formation in pollen tubes and incompatibility. *Biologia Pl.* **1**, 192.

Tupy, J. (1960). Sugar absorption, callose formation and the growth rate of pollen tubes. *Biologia Pl.* **2**, 169.

Ueda, Y., Minamide, T., Ogata, K., and Kamata, H. (1970). Lipids of fruits and vegetables and their physiological and qualitative role. 1. Method of quantitative determination of neutral lipids and changes of lipids during maturation of tomato fruit. *J. Food Sci. Technol. (Tokyo)* **17**, 49.

Vaks, B., and Lifshitz, A. (1976). An enzyme that attacks limonin at acid pH. *Lebensm-Wiss. t Technol.* **8**, 236.

Van Aller, R. T., and Nes, W. R. (1968). The phosphorylation of geraniol in germinating peas. *Phytochemistry* **7**, 85.

Van Buren, J. P. (1968). Adding calcium to snap beans at different stages in processing calcium uptake and texture of the canned product. *Food Technol. (Chicago)* **22**, 133.

Vance, C. P., and Sherwood, R. T. (1976). Regulation of lignin formation in reed canary grass in relation to disease resistance. *Plant Physiol.* **57**, 915.

Van Etten, G. H., Miller, R. W., Earle, F. R., Wolff, I. A., and Jones, Q. (1961). Hydroxyproline content of seed meals and distribution of the amino acid in kernel, seed coat, and pericarp. *J. Agric. Food Chem.* **9**, 433.

Van Teeling, C. G., Cansfield, P., and Gallop, R. A. (1971). An anthocyanin complex isolated from the syrup of canned blueberries. *J. Food Sci.* **36**, 1061.

Viani, R., Bricout, J., Marion, J. P., Muggler-Chavan, F., Reymond, D., and Egli, R. H. (1969). Sur la composition de l'arôme de tomate. *Helv. Chim. Acta* **52**, 887.

Vick, B. A., and Zimmerman, D. C. (1976). Lipoxygenase and hydroperoxide lyase in germinating watermelon seedlings. *Plant Physiol.* **57**, 580.

Virtanen, A. I. (1965). Studies on organic sulfur compounds and other labile substances in plants. *Phytochemistry* **4**, 207.

Virtanen, A. I., and Matikkala, E. J. (1959). The isolation of S-methyl cysteine sulfoxide and S-n-propyl-cysteine sulfoxide from onion (*Allium cepa*) and the antibiotic activity of crushed onion. *Acta Chem. Scand.* **13**, 1898.

Virtanen, A. I., and Spare, E. G. (1961). Isolation of the precursor of the lachrymatory factor in onion (*Allium cepa*), formation of the lachrymatory factor. *Suom. Kemistil. B* **35**, 28.

Virtanen, A. I., and Spare, E. G. (1962). On the enzymatic splitting of S-(1-propenyl)-cysteine sulfoxide and the formation of the lachrymatory factor. *suom. Kemistil. B* **35**, 28.

Von Loesecke, H. W. (1950). "Bananas." 2nd ed. Interscience Publishers, Inc., New York.

Wagenknecht, A. C., and Lee, F. A. (1956). The action of lipoxidase in frozen raw peas. *Food Res.* **21**, 605.

Wagenknecht, A. C., and Lee, F. A. (1958). Enzyme action and off-flavor in frozen peas. *Food Res.* **23**, 25.

Wagenknecht, A. C., Lee, F. A., and Boyle, F. P. (1952). The loss of chlorophyll in green peas during frozen storage and analysis. *Food Res.* **17**, 343.

Wagenknecht, A. C., Scheiner, D. M., and Van Buren, J. P. (1960). Anthocyanase activity and its possible relation to scald in sour cherries. *Food Technol. (Chicago)* **14**, 47.

Walker, G. C. (1964a). Color deterioration in frozen French beans (*Phaseolus vulgaris*). *J. Food Sci.* **29**, 383.

Walker, G. C. (1964b). Color deterioration in frozen French beans (*Phaseolus vulgaris*). 2. The effect of blanching. *J. Food Sci.* **29**, 389.

Walker, J. R. L. (1975). "The Biology of Plant Phenolics." Arnold, London.

Wallner, S. J., and Bloom, H. L. (1977). Characteristics of tomato cell-wall degradation *in vivo*. *Plant Physiol.* **60**, 207.

Wallner, S. J., and Walker, J. E. (1975). Glycosidases in cell-wall degrading extracts of ripening tomato fruits. *Plant Physiol.* **55**, 94.

Walton, E., and Butt, V. S. (1970). The activation of cinnamate by an enzyme from leaves of spinach beet (*Beta vulgaris* L.ssp vulgaris). *J. Exp. Bot.* **21**, 887.

Walton, E., and Butt, V. S. (1971). Demonstration of cinnamyl CoA synthetase activity in leaf extracts. *Phytochemistry* **10**, 295.

Wardrop, A. B., and Bland, D. E. (1958). The process of lignification in woody plants. In Proc. 4th Int. Congr. Biochem. Vienna Vol. 2, pp. 92–116. (K. Krantzl and G. Billek).

Weast, C. A., and Mackinney, G. (1940). Chlorophyllase. *J. Biol. Chem.* **133**, 551.

Webster, D. B., and Currier, H. B. (1965). Callose: Lateral movement of assimilates from phloem. *Science* **150**, 1610.

Webster, D. B., and Currier, H. B. (1968). Heat induced callose and lateral movement of assimilates from phloem. *Can. J. Bot.* **46**, 1215.

Weissenbock, G. (1975). Aktivatsverlauf der phenylalanin-Tryosin-Ammonium-Lyase (PAL,TAL) und chalkors-Flavonon-Isomerse in Vergleich zur C-Glycoxylflavan-Akkumulation in wachsenden Hafersprss (*Avena Sativa* L.) bei Belichtung und Dunkelheit. Z. *Pflanzenphysiol.* **74**, 226.

Weurman, C. (1954). Pectase in Doyenne Bousooch pears and changes in the quality of the enzyme during development. *Acta Bot. Neerl.* **3**, 100.

White, R. C., Jones, I. D., and Gibbs, E. (1963). Determination of chlorophylls, chlorophyllides, pheophytins and pheophorbides in plant material. *J. Food Sci.* **28**, 431.

Whitmore, F. M. (1971). Effect of indoleacetic acid and hydroxyproline on isoenzymes of peroxidases in wheat coleoptiles. *Plant Physiol.* **47**, 169.

Whitenberger, R. T., and Hills, C. H. (1953). Studies on the processing of red cherries. 1. Changes in fresh red cherries caused by bruising, cooling and soaking. *Food Technol. (Chicago)* **1**, 29.

Wick, E. L., Yamanishi, T., Kobayashi, A., Valenzuela, S., and Issenberg, P. (1969). Volatile constituents of banana (*M. cavendishii*, variety Valery) *J. Agric. Food Chem.* **17**, 751.

Wiley, R. C., and Lee, Y. S. (1970). Modifying texture of processed apple slices. *Food Technol. (Chicago)* **24**, 1169.

Wiley, R. C., Elehmony, N., Kramer, A., and Hager, F. J. (1956). The shear press—an instrument for measuring the quality of foods. Application to asparagus. *Food Technol. (Chicago)* **10**, 439.

Wiley, R. C., Meschter, E. E., Lacey, H., and Kramer, A. (1963). Use of the shear press in determining fibrousness of raw and canned asparagus. *Food Technol. (Chicago)* **17**, 81.

Willstater, R., and Hocheder, F. (1907). Uber die Einwirkung von Sauron. *Justus Liebigs. Ann. Chem.* **354**, 205.

Wilson, D. G., King, K. W., and Burns, R. H. (1954). Transamination reactions in plants. *J. Biol. Chem.* **208**, 863.

Winterstein, A., Studer, A., and Rüegg, R. (1960). Neuere Eigebnisse der Carotinoid-forschung. *Chem. Ber.* **93**, 2951.

Winter, M., and Gautschi, F. (1962). Odeur et constitution. XX. Synthèses des *cis*-et *trans*-hexene-3-al. *Helv. Chim. Acta* **45**, 2567.

Witting, L. A., and Porter, J. W. (1959). Intermediates in the conversion of mevalonic acid by a rat-liver enzyme system. *J. Biol. Chem.* **234**, 2841.

Wong, E., and Moustafa, E. (1966). Flavonone biosynthesis. *Tetrahedron Lett.* **26**, 3021.

Wong, W. J. L., and ap Rees, T. (1971). Carbohydrate oxidation in shell and cortex isolated from roots of *Pisum sativum*. *Biochim. Biophys. Acta* **252**, 296.

Worth, H. G. J. (1967). The chemistry and biochemistry of pectic substances. *Chem. Rev.* **67**, 465.

Wright, D., Brown, S. A., and Neish, A. C. (1958). Studies of lignin biosynthesis using isotopic carbon. VI. Formation of the side chain of the phenylpropane monomer. *Can. J. Biochem. Physiol.* **36**, 1037.

Wrolstad, R. E., Putnam, T. P., and Varseveld, G. W. (1970). Color quality of frozen strawberries. Effect of anthocyanin, pH, total acidity and ascorbic acid variability. *J. Food Sci.* **35**, 448.

Yang, H. Y., and Steele, W. F. (1958). Removal of excessive anthocyanin pigment by enzymes. *Food Technol. (Chicago)* **12**, 517.

Yasumoto, K., Iwami, K., and Mitsuda, H. (1971a). A new sulfur-containing peptide from Lentinus ediodes acting as a precursor for lenthionine. *Agric. Biol. Chem.* **35**, 2059.

Yasumoto, K., Iwami, K., and Mitsuda, H. (1971b). Enzyme catalysed evolution of lenth-ionine from lentinic acid. *Agric. Biol. Chem.* **35**, 2070.

Yoshida, S. (1969). Biosynthesis and conversion of aromatic amino-acids in plants. *Annu. Rev. Plant Physiol.* **20**, 41.

Younathan, M. T., and Watts, B. M. (1959). Relationship of meat pigments to lipid oxida-tion. *Food Res.* **24**, 728.

Young, R. (1972). Relation of ethylene and cellulase activity to abscission of freeze-injured citrus fruit. *Proc. Am. Soc. Hortic. Sci.* **97**, 133.

Yu, M. H., and Spencer, M. (1969). Conversion of L-leucine to certain Keto acids by a tomato enzyme preparation. *Phytochemistry* **8**, 1173.

Yu, M. H., and Spencer, M. (1970). α-Alanine aminotransferase from tomato fruit. *Phytochemistry* **9**, 341.

Yu, M. H., Olson, L. E., and Salunkhe, D. K. (1967). Precursors of volatile components in tomato fruit. I. Compositional changes during development. *Phytochemistry* **6**, 1457.

Yu, M. H., Salunkhe, D. K., and Olson, L. E. (1968a). Production of 3-methylbutanol from L-leucine by tomato extract. *Plant Cell Physiol.* **9**, 633.

Yu, M. H., Olson, L. E., and Salunkhe, D. K. (1968b). Precursors of volatile components in tomato fruit. II. Enzymatic production of carbonyl compounds. *Phytochemistry* **7**, 555.

Yu, M. H., Olson, L. E., and Salunkhe, D. K. (1968c). Precursors of volatile components in tomato fruit. III. Enzymatic reaction products. *Phytochemistry* **7**, 561.

Zaubermann, A., and Schiffman-Nadel, M. (1972). Pectin methylesterase and polygalacturanase in avocado fruit at various stages of development. *Plant Physiol.* **49**, 864.

Zechmeister, L., and Tuzson, P. (1939). Reversible isomerization of carotenoids by iodine catalysis. *Ber. Dtsch. Chem. Ges. B* **72**, 1340; cf. *Chem. Abstr.* **33**, 7727 (1939).

Zenk, M. H., and Gross, G. G. (1974). The enzymic reduction of cinnamic acids. *Recent Adv. Phytochem.* **8**, 88.

Zimmerman, D. C. (1966). A new product of linoleic acid oxidation by a flaxseed enzyme. *Biochem. Biophys. Res. Commun.* **23**, 398.

Zimmerman, D. C., and Vick, A. A. (1970). Hydroperoxide isomerase. A new enzyme of lipid metabolism. *Plant Physiol.* **46**, 445.

Index

I

J

K

L

M

N

O

P

FOOD SCIENCE AND TECHNOLOGY

A SERIES OF MONOGRAPHS

Maynard A. Amerine, Rose Marie Pangborn, and Edward B. Roessler, PRINCIPLES OF SENSORY EVALUATION OF FOOD. 1965.

S. M. Herschdoerfer, QUALITY CONTROL IN THE FOOD INDUSTRY. Volume I – 1967. Volume II – 1968. Volume III – 1972.

Hans Riemann, FOOD-BORNE INFECTIONS AND INTOXICATIONS. 1969.

Irvin E. Liener, TOXIC CONSTITUENTS OF PLANT FOODSTUFFS. 1969.

Martin Glicksman, GUM TECHNOLOGY IN THE FOOD INDUSTRY. 1970.

L. A. Goldblatt, AFLATOXIN. 1970.

Maynard A. Joslyn, METHODS IN FOOD ANALYSIS, second edition. 1970.

A. C. Hulme (ed.), THE BIOCHEMISTRY OF FRUITS AND THEIR PRODUCTS. Volume 1 – 1970. Volume 2 – 1971.

G. Ohloff and A. F. Thomas, GUSTATION AND OLFACTION. 1971.

George F. Stewart and Maynard A. Amerine, INTRODUCTION TO FOOD SCIENCE AND TECHNOLOGY. 1973.

C. R. Stumbo, THERMOBACTERIOLOGY IN FOOD PROCESSING, second edition. 1973.

Irvin E. Liener (ed.), TOXIC CONSTITUENTS OF ANIMAL FOODSTUFFS, second edition. 1974.

Aaron M. Altschul (ed.), NEW PROTEIN FOODS: Volume 1, TECHNOLOGY, PART A – 1974. Volume 2, TECHNOLOGY, PART B — 1976. Volume 3, ANIMAL PROTEIN SUPPLIES, PART A — 1978.

S. A. Goldblith, L. Rey, and W. W. Rothmayr, FREEZE DRYING AND ADVANCED FOOD TECHNOLOGY. 1975.

R. B. Duckworth (ed.), WATER RELATIONS OF FOOD. 1975.

Gerald Reed (ed.), ENZYMES IN FOOD PROCESSING, second edition. 1975.

A. G. Ward and A. Courts (eds.), THE SCIENCE AND TECHNOLOGY OF GELATIN. 1976.

John A. Troller and J. H. B. Christian, WATER ACTIVITY AND FOOD. 1978.

A. E. Bender, FOOD PROCESSING AND NUTRITION. 1978.

D. R. Osborne and P. Voogt, THE ANALYSIS OF NUTRIENTS IN FOODS. 1978.

Marcel Loncin and R. L. Merson, FOOD ENGINEERING: PRINCIPLES AND SELECTED APPLICATIONS. 1979.

Hans Riemann and Frank L. Bryan (eds.), FOOD-BORNE INFECTIONS AND INTOXICATIONS, Second Edition. 1979.

N. A. Michael Eskin, PLANT PIGMENTS, FLAVORS AND TEXTURES: THE CHEMISTRY AND BIOCHEMISTRY OF SELECTED COMPOUNDS. 1979.

J. G. Vaughan (ed.), FOOD MICROSCOPY. 1979.

J. R. A. Pollock (ed.), BREWING SCIENCE. Volume 1. 1979.

In preparation

Irvin E. Liener (ed.), TOXIC CONSTITUENTS OF PLANT FOODSTUFFS, Second Edition. 1979.